TRANSPARENT CONDUCTING FILMS

Edited by **Kaushik Pal**

Transparent Conducting Films
http://dx.doi.org/10.5772/intechopen.72957
Edited by Kaushik Pal

Contributors

Xiaogang Sun, Manyuan Cai, Long Chen, Zhiwen Qiu, Jie Wang, Xu Li, Wei Chen, Yapan Huang, Chengcheng Wei, Hao Hu, Guodong Liang, Iskandar Yahya, Seri Mastura Mustaza, Huda Abdullah, Wan Wardatul Amani Wan Salim, Nur Alya Batrisya Ismail, Nurul Izzati Ramli, Firdaus Abd-Wahab, Joong Tark Han, Prof.(Dr.) Kaushik Pal

Notice

Statements and opinions expressed in the chapters are these of the individual contributors and not necessarily those of the editors or publisher. No responsibility is accepted for the accuracy of information contained in the published chapters. The publisher assumes no responsibility for any damage or injury to persons or property arising out of the use of any materials, instructions, methods or ideas contained in the book.

First published in London, United Kingdom, 2019 by IntechOpen
IntechOpen is the global imprint of INTECHOPEN LIMITED, registered in England and Wales, registration number: 11086078, The Shard, 25th floor, 32 London Bridge Street
London, SE19SG – United Kingdom
Printed in Croatia

British Library Cataloguing-in-Publication Data
A catalogue record for this book is available from the British Library

Additional hard and PDF copies can be obtained from orders@intechopen.com

Transparent Conducting Films, Edited by Kaushik Pal
p. cm.
Print ISBN 978-1-83880-494-7
Online ISBN 978-1-83880-495-4
eBook (PDF) ISBN 978-1-83880-987-4

We are IntechOpen,
the world's leading publisher of
Open Access books
Built by scientists, for scientists

4,200+
Open access books available

116,000+
International authors and editors

125M+
Downloads

Our authors are among the

151
Countries delivered to

Top 1%
most cited scientists

12.2%
Contributors from top 500 universities

CLARIVATE ANALYTICS
BOOK
CITATION
INDEX
INDEXED

WEB OF SCIENCE™

Selection of our books indexed in the Book Citation Index
in Web of Science™ Core Collection (BKCI)

Interested in publishing with us?
Contact book.department@intechopen.com

Numbers displayed above are based on latest data collected.
For more information visit www.intechopen.com

Meet the editor

Prof. (Dr.) Kaushik Pal was born in India and received his PhD from the University of Kalyani (India). He has received many prestigious awards: Marie-Curie Experienced Researcher (Postdoctoral Fellowship) offered by the European Commission in Greece; Brain Korea National Research Foundation Visiting Scientist Fellowship in South Korea; Senior Postdoctoral Fellow at Wuhan University, China; and Chief-Scientist & Faculty CAS Fellow by the Chinese Academy of Science. He is now working as a research professor, group leader, and independent scientist at the Department of Nanotechnology, Bharath University (BIHER), Chennai. He was invited as a visiting professor in March 2019 at University Technology Malaysia and University of Malaya, Kuala Lumpur, Malaysia. He is acting editor-in-chief of international peer-reviewed journals from publishers such as Pan Standford, Elsevier, and Springer. He has edited and contributed to significant numbers of book chapters (15) and review articles (70) and has reviewed 95 research articles. Prof. Pal has organized and has been the chairperson for around 25 national/international events/symposiums/conferences/workshops and has contributed to around 8 plenary, 25 keynote, and 30 invited lectures. He is distinguished in the worldwide nanotechnology and materials research community.

Contents

Preface

Transparent conducting film (TCF) is a unique class of designing process that exhibits transparency and electronic conductivity simultaneously. The novelty of research has widespread utilization in displays, photovoltaics, low-e windows, optoelectronics, nanoelectronics, and flexible electronics.

In many aspects of these applications, TCFs are used in their role as transparent contacts. However, increasingly, the demands required have extended beyond the combination of conductivity and transparency, where higher performance is needed, and now includes work function, synthesis, structural morphology, designing processes and patterning requirements, long-term stability, cost-effectiveness, and elemental abundance/green nanomaterials. As these needs began to emerge over the last 5 years, they have stimulated a dramatic resurgence of research in the field leading to many new materials and processes. The purpose of this book is both to provide a snapshot of the unique and enabling work in the field and to provide indications of what might be coming over the next few decades.

Over the past 5–10 years, the field has exploded to include a vastly increased number of n-type materials and a class of new p-type materials. In addition, the historically held view that crystalline materials have superior properties has been challenged by the emergence of new materials-coating TCFs that have properties as good as or better than their potential application counterparts.

These materials have led to the development of amorphous oxide transistors, which offer the advantage of low-temperature processing and the promise of flexible electronics on polymer substrates. In their role as a channel material in thin film transistor structures, transparent conducting oxides with controlled carrier densities are often termed transparent oxide semiconductors since their key properties may lie in the limited to nonconductive regime. We have organized the book to capture this diversity of materials, processes, and applications. Over the next few years, we expect these materials will become increasingly important for TCF techniques. Their inclusion in this volume at present is, however, beyond its intended scope.

Professor (Dr.) Kaushik Pal
Bharath University, India

Prosperity for the Commercialisation of "Transparent Conducting Films"

Chapter 1

Introductory Chapter: Transparent Conducting Films

Kaushik Pal

Additional information is available at the end of the chapter

http://dx.doi.org/10.5772/intechopen.85577

1. Implementation and benefits of "Transparent Conducting Films"

There has been an increasing demand for functional films, which combine a film substrate with various features. However, the main research goal of the transparent conducting films (TCF) and materials has been rapidly promising to scientists as well as industries. This continuing transformation is taking place at all levels: technologies, applications, developers and suppliers. Owing to their processability, stability, and high conductivity, carbon nanotubes has received significant attention from electronics-industry researchers over the past several years as an alternative to ITO. As per current trends for transparent conductive films increases, transparent electrode materials alternatives to ITO and active research and development for commercialization of such materials are being conducted. Meanwhile, transparent conductive films that have conductivity while being transparent are heavily used as essential elements for touch panels of smartphones or tablets or transparent electrodes of solar cells or other products. In particular entitled book "Transparent Conducting Films", we provide the most comprehensive and authoritative chapters are based upon years of research as we have been tracking and analyzing TCF industry since 2008. Those useful chapters are listed below in contained book;

- Carbon nanotube transparent conducting film

- Carbon nanotube activated thin-film transparent conductor applications

- Conductive polymers in biosensors

- Nanocarbon-based transparent conducting films

Our expert team of reviewers and editors has also been independently analyzed and peer reviewed those individual articles to flourish emerging target applications. Indeed, most of the articles, particularly concentrated on OLED lighting, wearable technology, in-mold electronics, smart windows, OPVs, DSSCs, perovskites, and touch screens. This enables us to assess the market from an application as well as technology point of view. The approach mainly the author used for fabrication is highly reproducible and creates a chemically stable configuration with a tunable tradeoff between transparency and conductive properties. In the new study, the contributors used an approach called colloidal lithography to create transparent conductive silver thin films.

If researchers would like to get specific knowledge on this topic from the beginning, the best advice would be to choose firstly the branch among an incomprehensible canopy of transparent conducting films and its various applied studies [1, 2]. The book aimed to show how the field is studied in different countries and what is common for all spectroscopic or microscopic investigations. The results from these experimental studies are very important outcomes of model experiments carried out on cultivating thin film techniques.

The phase, purity, stability and morphology of the composite and its constitutes have been also analyzed in those chapters. Hence, it possesses superior thermal properties and higher thermal stabilities of its layers [3, 4], qualifying it to be used in various thermo-electric devices [5] and photovoltaics. Indeed, the optical properties can be studied by utilizing optical absorption spectrum calculated optical energy band gap of the conducting film [6, 7]. The electrical parameters such as dielectric constant, tangent loss, AC conductivity as a function of frequency with fixed typical temperature also analyze.

The overall studies and investigated results in our individual chapter. Through the entire book in this year will get scope to learn more about the market trends, the key questions, latest prices, novelty of applications, e.g., transparent electrodes, flexible displays or wearable devices, OPV (organic photovoltaics) cells, light control glasses or films, organic EL lighting, transparent antennas, transparent electric wave shielding materials, and fine-tuned our analysis, insight and forecasts to reflect the latest research.

We also believe that it will be most help beginner research scholars, scientists, academicians in current understanding and advise them quite novel and non-standard approaches to find and decipher the mechanisms of transparent conducting film methodology and its application.

Finally, we would like to thank all the concern authors for their endless contributions and hard work to match and unify the "philosophy" of this book. We also thank to our colleagues from University Federal Rio de Jenerio, Brazil and Mahatma Gandhi University, Kerala, India who supported us and helped us in preparation and edition of the chapters, especially to those who raised complex questions and promoted us to answer them. We are personally very grateful to the "In-Tech" Publisher, especially Ms. Anita Condic, who assisted us in the arrangement of the book and scheduling our activities.

Author details

Kaushik Pal

Address all correspondence to: kaushikphysics@gmail.com

Department of Nanotechnology, Bharath Institute of Higher Education and Research, Bharath University, Chennai, Tamil Nadu, India

References

[1] Rakesh AA. Transparent conducting oxide films forvarious applications: A review. Reviews on Advanced Materials Science. 2018;**53**:79-89

[2] Mizoguchi H, Woodward PM. Electronic structure studies of main group oxides possessing edge-sharing octahedra: Implications for the design of transparent conducting oxides. Chemistry of Materials. 2004;**16**(16):5233-5248

[3] Pal K, Maiti UN, Majumder TP, Dash P, Mishra NC, Bennis N, et al. Ultraviolet visible spectroscopy of CdS nano-wires doped ferroelectric liquid crystal. Journal of Molecular Liquids. 2011;**164**:233-238

[4] Pal K, Majumder TP, Neogy C, Debnath SC. Optical, dielectric and microscopic observation of different phases TiO_2 metal host nanowires. Journal of Molecular Structure. 2012;**1016**:30-38

[5] Pal K, Maiti UN, Majumder TP, Debnath SC, Ghosh S, Roy SK, et al. Switching of ferroelectric liquid crystal doped with cetyltrimethyl ammonium bromide assisted CdS nanostructures. Nanotechnology. 2013;**24**:125702

[6] Sagadevan S, Das I, Pal K, Murugasen P, Singh P. Optical and electrical smart response of chemically stabilized graphene oxide. Journal of Materials Science: Materials in Electronics. 2017;**28**(7):5235-5243

[7] Pal K, Mohan MLNM, Foley M, Ahmed W. Emerging assembly of ZnO-nanowires/graphene dispersed liquid crystal for switchable device modulation. Organic Electronics. 2018;**56**:291-304

Novel Growth Mechanism of Nanocarbon for Transparent Conducting Films Utilization

Chapter 2

Interfacial Engineering of Flexible Transparent Conducting Films

Joong Tark Han and Geon-Woong Lee

Additional information is available at the end of the chapter

http://dx.doi.org/10.5772/intechopen.80259

Abstract

One-dimensional (1D) carbon nanotubes (CNTs) and silver nanowires (AgNWs) have been used as replacements for brittle indium tin oxide (ITO) in the fabrication of transparent conducting films (TCFs), which can be used in opto-electronic devices such as screen panels, solar cell panels, and organic light-emitting diodes. This chapter describes a fabrication method of high-performance TCFs by solution processing of single-walled CNTs (SWCNTs) and AgNWs. Highly uniform TCFs with SWCNTs and AgNW inks were fabricated using spray deposition. Their performance was modulated by interfacial engineering such as overcoating with silane compound for densification of SWCNT networks and chemical or photothermal welding of SWCNT networks on thermoplastic substrates. Moreover, the hybridization of SWCNTs, AgNWs, and graphene oxide nanosheets is a promising approach to mitigate their drawbacks via p-type doping, electrical stabilization, or interfacial stabilization on plastic substrates. The rational control of 1D material networks can provide a good opportunity to fabricate high-performance TCFs for flexible opto-electronic devices.

Keywords: single-walled carbon nanotubes, silver nanowires, interfacial engineering, graphene oxide, dispersion, sheet resistance

1. Introduction

One-dimensional (1D) conducting nanomaterials such as carbon nanotubes (CNTs) and metal nanowires have been studied to replace brittle indium tin oxide (ITO) films for flexible opto-electronic devices because of their flexibility and high electrical conductivity as well as solution processability [1–5]. There are growing needs for high-performance transparent conducting films (TCFs) with flexibility to realize flexible displays or solar cells. Solution processing of conducting nanomaterials for TCFs has many challenging issues in order to

Silver nanowire (AgNW)

Single-walled carbon nanotubes (SWCNTs)

1D/1D hybrid TCF

SWCNT or Silver nanowire (AgNW)

Graphene oxide
Reduced graphene oxide

1D/2D hybrid TCF

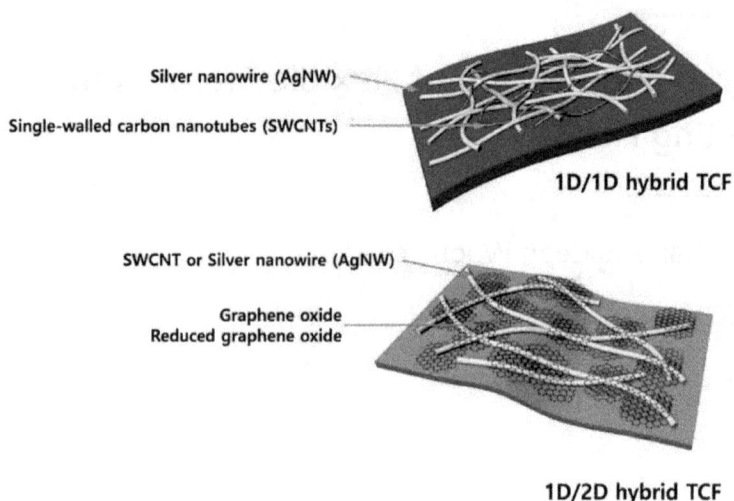

Figure 1. Scheme of hybrid TCFs fabricated with 1D/1D hybrid materials and 1D/2D hybrid materials by solution processing.

achieve high performance, including the intrinsic properties of the materials, the dispersion of nanomaterials, and interfacial engineering of coating films on plastic substrates. Moreover, to mitigate the drawbacks of each conducting nanomaterial, we need a rational hybridization strategy to achieve the fabrication of high performance TCFs on plastic substrates (**Figure 1**).

Therefore, this chapter describes some of the research on the fabrication of high-performance TCFs based on single-walled CNTs (SWCNTs) and silver nanowires (AgNWs) over the past 8 years that addresses these and other challenges, with an emphasis on our own efforts. We begin with the realization of TCFs with high uniformity by spray deposition and then describe the interfacial engineering of TCFs on plastic substrates. Furthermore, we describe the fabrication of flexible TCFs with 1D/1D hybrid structures and 1D/2D hybrid materials with SWCNTs and AgNWs as 1D materials and graphene oxide as a 2D material. We conclude with some discussion of future directions and the remaining challenges in chemically exfoliated graphene technologies.

2. Fabrication of TCFs by spray coating

Spray coating methods can be used to fabricate flexible TCFs with aqueous single-walled carbon nanotube (SWCNT) solutions or silver nanowire (AgNW) solutions on plastic substrates. As shown in **Figure 2**, thin films were deposited on the substrate by the atomization of aqueous solution using high-pressure nitrogen gas through a spray nozzle. The gas flow rate, nozzle height, and pitch should be controlled to fabricate uniform films with high opto-electrical performance. As a model system, SWCNT solution dispersed in aqueous surfactant and aqueous AgNW solution containing a small amount (0.01 wt%) of polyvinylpyrrolidone (PVP) were used to investigate the spreading behavior on surface energy-controlled substrates. To control the surface energy of the substrate, plastic substrates were irradiated with UV-ozone (UVO). The wettability of coating inks is critical for fabrication of uniform films by spraying. **Figure 3**

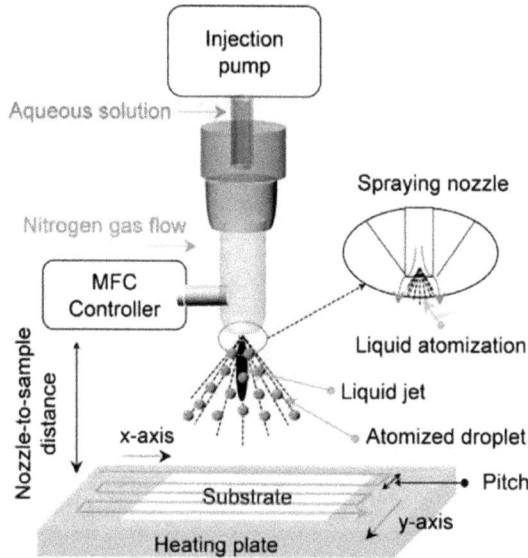

Figure 2. Schematic of the automatic spray coating system with mass flow controller, injection pump, and atomizing nozzle. The X- and Y-direction can be controlledautomatically by robotics [6].

shows the contact angle (CA) change with an increase in UVO irradiation time of polycarbonate substrates. The CA of the SWCNT/surfactant solution decreases from 15 to 10°. The CA of the aqueous AgNW solution decreases from 68.7 to 36.6° with an increasing UVO irradiation time, and the size of the deposited liquid droplet increases from 8 to 13 mm. The nozzle height and the spraying pitch were optimized to 70 and 7 mm, respectively.

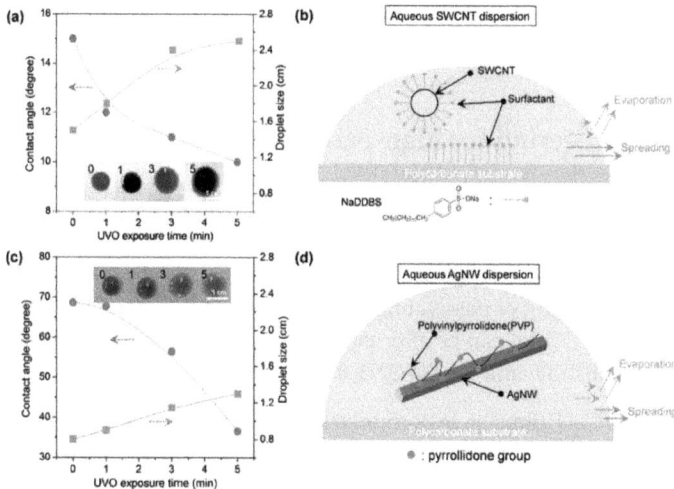

Figure 3. Contact angles and spread droplet sizes of (a) the aqueous SWCNT solution dispersed by sodium dodecylbenzene sulfonate and (c) the aqueous AgNW solution containing PVP on polycarbonate substratesby varying the UVO exposure time. The inset photoimages in (a) and (c) show the spread SWCNT and AgNW droplet sizes on the substrate by varying the UVO-irradiation time indicated by values. (b) and (d) Schematics of spreading of the SWCNT (b) and AgNW (d) droplets on substrates [6].

SWCNT films		
On pristine PC (7.2%)	On UVO-treated PC (7.0%)	HOGO-coated (6.4%)
(a)	(b)	(c)

AgNW films		
On pristine PC (17.2%)	On UVO-treated PC (10.5%)	HOGO-coated (7.2%)
(d)	(e)	(f)

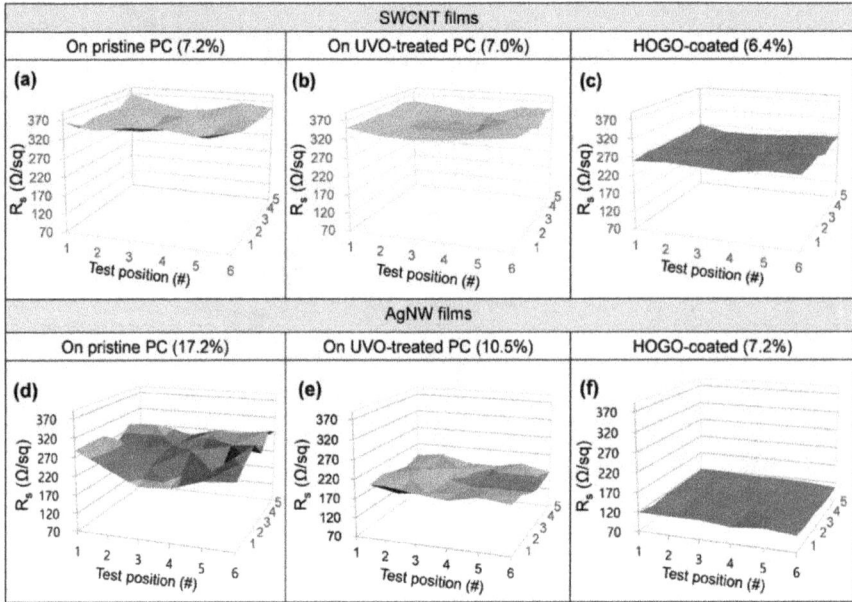

Figure 4. The sheet resistance (R_s) uniformity of (a–c) the SWCNT films and (d–f) the AgNW films on (a, d) pristine polycarbonate (PC), (b, e) UVO-irradiated PC substrates, and (c, f) after graphene oxide (HOGO) coating of the conducting films fabricated on UVO-treated substrates [6].

Another way to enhance the uniformity of the films is deposition of hydrophilic graphene oxide (GO) nanosheets onto the substrate. **Figure 4** shows the sheet resistance (R_s) distribution of the SWCNT and the AgNW films spray-coated on surface energy-controlled substrates and after deposition of GO nanosheets onto the films. After UVO treatment, the R_s uniformity of AgNW films was dramatically improved and reached 7.2%, resulting in $T = 98\%$ and $R_s = 100$ Ω/sq for the highlyoxidized GO (HOGO)-coated AgNW films.

3. Interfacial engineering for high-performance TCFs

3.1. Modulation of the sheet resistance of SWCNT-based TCFs by silane sol

In this study, we investigated the effect of the interfacial tension between bare SWCNT network films and a top-coating of passivation materials on the R_s of the film. We demonstrated that the R_s of the SWCNT film can be affected by a thermal expansion coefficient (CTE) mismatch between the substrate and the SWCNT film.

The spray-coated SWCNT films have porous structures on a scale of tens of nanometers. The R_s and transmittance are related by [7].

$$T(\lambda) = \left(1 + \frac{188.5}{R_s}\frac{\sigma_{Op}(\lambda)}{\sigma_{DC}}\right)^{-2}, \tag{1}$$

where σ_{DC} and σ_{Op} are the DC and optical conductivities, respectively.

The conductivity, σ_{DC}, of the disordered nanotube films depends on the number density of the network junctions, N_f, which in turn scales with the network morphology through the film fill-factor, V_f, the mean diameter, $<D>$, of the bundles, and the mean junction resistance, $<R_j>$ [8–11],

$$\sigma_{DC} = \frac{K}{< R_j > < D >^3} V_f^2 \qquad (2)$$

Here, K is the proportionality factor that scales with the bundle length. Therefore, if we can reduce $<R_j>$ and V_f, the sheet resistance of the SWCNT films can be improved. To realize this, the SWCNT films were coated with silane sols by considering their surface energy. Considering the interfacial tension between the SWCNT film and silane sols, two top-coating materials such as a tetraorthosilicate (TEOS) sol with silanol groups and methyltrimethoxysilane (MTMS) sol with hydrophobic methyl groups were used. It is worth noting that top-coating with TEOS sol unexpectedly decreased the R_s of the film to less than 80% of the R_s of the as-prepared film. However, the R_s values of MTMS sol-coated SWCNT films gradually increased. This large disparity between MTMS and TEOS sols can be explained by a change in the contact resistance between the bundles. Hydrophilic TEOS sol can densify the hydrophobic SWCNT networks, while MTMS sol, having methyl groups, can penetrate the hydrophobic SWCNT networks, resulting in an increase of the contact resistance of SWCNTs. This interfacial tension effect was minimized by deposition of gold chloride solution onto the SWCNT film (**Figure 5b**) to make it hydrophilic, as shown in **Figure 5c**.

Figure 6 shows the R_s change after heating at 130°C and cooling. Bare PET, hard-coated PET, and glass substrates were used to illustrate the CTE mismatch effect on the R_s changes of SWCNT films. Interestingly, the R_s of the SWCNTs on bare PET substrates increased by 40% relative to the initial values, while the R_s increase was suppressed in bare SWCNT films on hard-coated PET and glass. These results imply that the CTE value should be considered in order to obtain highly stable SWCNT TCFs on plastic substrates. To illustrate this phenomenon, a Raman spectroscopic study was performed, and the G+ and G− peak positions related to the strain of SWCNTs were compared. The G-band frequencies for SWCNT films on bare PET were up-shifted by 1–2 cm⁻¹ after heating at 130°C and cooling, which corresponds to a

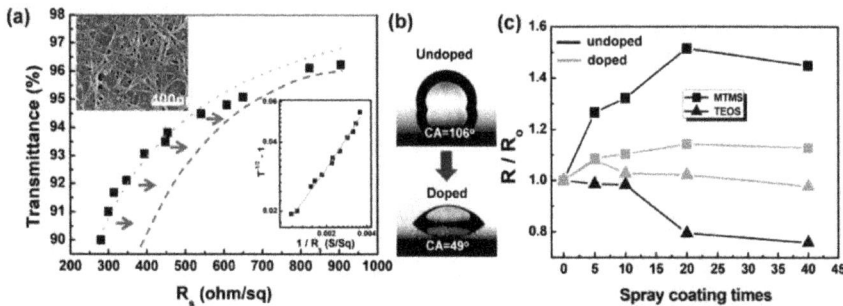

Figure 5. (a) The R_s versus transmittance plot of SWCNT film deposited by spraying on PET substrates. (b) Wettability of pristine SWCNT film and SWCNT film doped with gold chloride. (c) The R_s change of pristine and doped SWCNT films by varying spray coating times of top-coating materials (methyl trimethoxysilane (MTMS) sol, tetraethoxysilane (TEOS) sol) after baking at 80°C for 1 h [12].

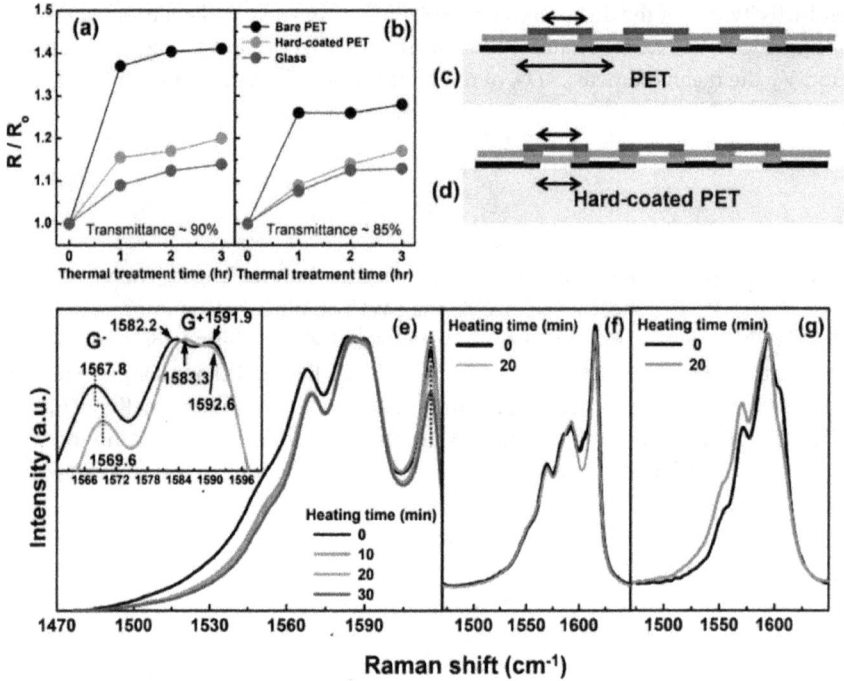

Figure 6. (a, b) The R_s changes of SWCNT films with different transmittance values, after heating at 130°C, as a function of thermal treatment time. (c, d) Scheme of thermal expansion mismatch between the SWCNT layers and bare PET or hard-coated PET after heating and cooling. (e–g) Raman spectra (G band) of SWCNT films fabricated on (e) bare PET, (f) hard-coated PET, and (g) glass after heating at 130°C for 20 min, followed by cooling [12].

compressive strain of ~0.1%. This compressive strain may cause the increase of the R_s of the SWCNT film on bare PET.

3.2. Self-passivation of SWCNT films on plastic substrates by nanowelding

Plastic substrates are generally used to fabricate flexible TCFs by deposition of CNTs or metal nanowires. In particular, the electrical properties of SWCNT network films are sensitive to humidity and temperature. In this context, top-coating with passivation materials or hybridization with binder materials are applicable for improving the stability of TCFs. Another way to passivate TCFs is welding or embedding in plastic substrates by chemical or thermal treatments. **Figure 7** shows the R_s change of the SWCNT films after deposition of solvents. To investigate the solvent effects, we used solvents with optimal polarity and affinity for the PET substrate. Moreover, the presence of electron-donating and electron-withdrawing groups in the solvent molecules can affect the electronic structure of the SWCNTs. Thus, nonpolar solvents were selected. In particular, aromatic hydrocarbon, benzene, and toluene can swell the PET substrate. Most interestingly, deposition of toluene or benzene decreased the R_s of the SWCNT films. After doping with gold chloride, the R_s and transmittance of the film were measured to be 85 Ω/sq and 90%, respectively. Moreover, I-V plots measured after solvent

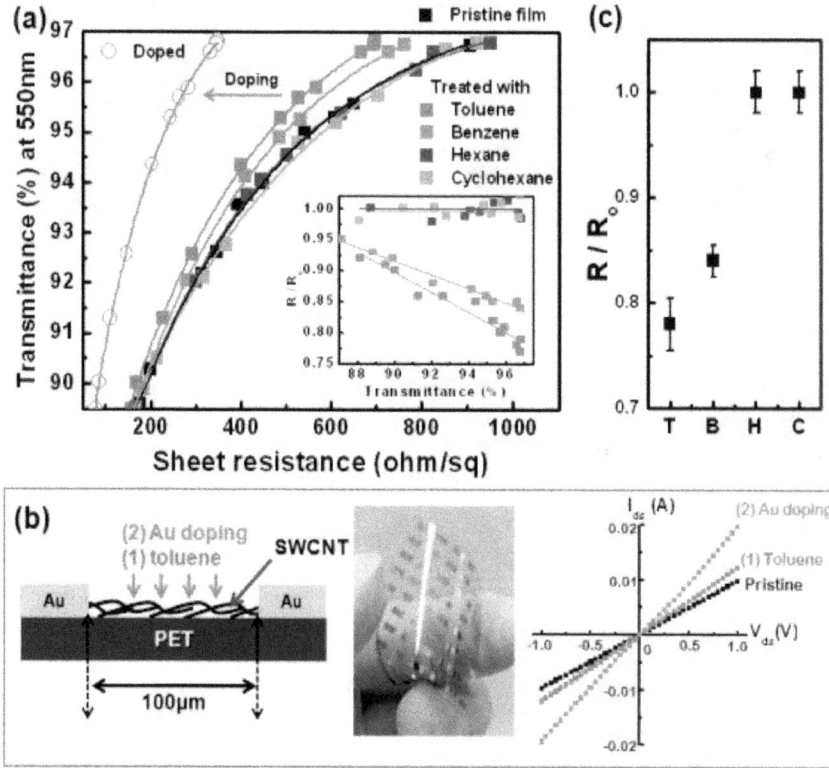

Figure 7. (a) R_s versus transmittance plots for pristine SWCNT films prepared from a SWNT/SDBS solution on PET surfaces, and after deposition of solvents and dopants. (b) In-situ conductivity measurements of SWCNT films after deposition of toluene and gold chloride. (c) R_s changes of bare SWCNT films in comparison with the same films treated with solvents: toluene (T), benzene (B), hexane (H), and cyclohexane (C) [13].

deposition show clearly that the electrical conductivity of the SWCNT films was enhanced after toluene deposition, which can swell PET substrates.

Figure 8 shows that large SWCNT bundles were welded, and small bundles were embedded on the PET substrate after spraying aromatic hydrocarbons, while spraying cyclohexane did not trigger welding. The strain induced on the SWCNT networks during network formation on the substrate may cause an initial high resistance in the SWCNT network film. Thus, solvent-induced chemical welding of the SWCNT film can release their strain. The recovery of the G band in the Raman spectra of the SWCNT films demonstrates strain relaxation via chemical welding.

Thermal treatment is an alternative way to produce SWCNT film-substrate welding without any chemicals. In particular, fast selective heating of CNTs on plastic substrates can provide an interesting opportunity for thermal welding [14, 15]. Microwaves irradiate the SWCNT films inside the rectangular waveguide microwave applicator, within which the microwave electric field is well defined and controlled. The microwave mode in the applicator is a

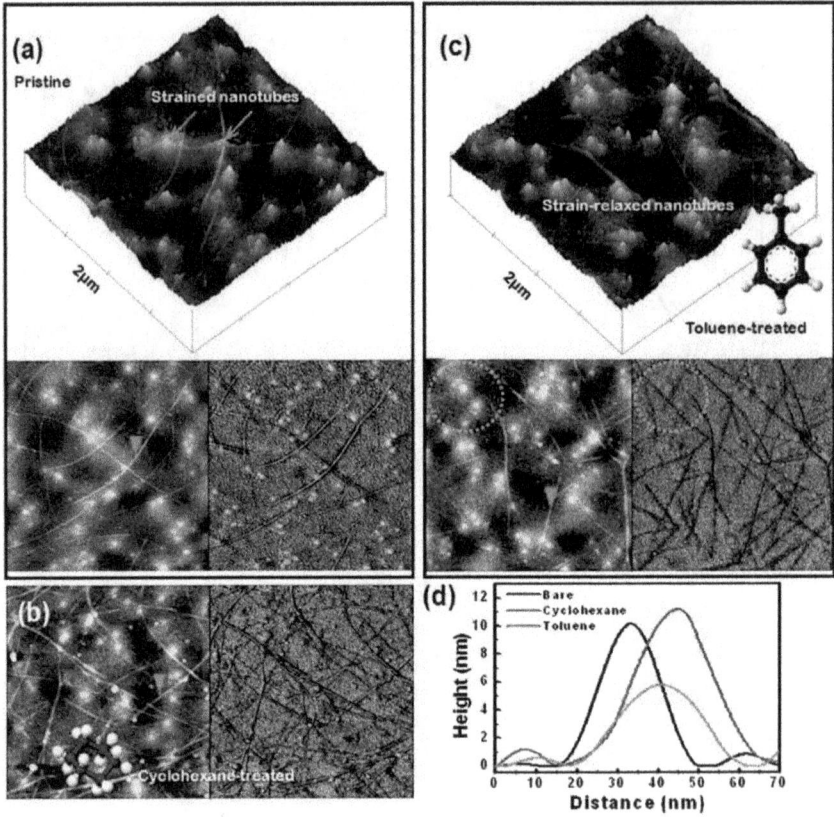

Figure 8. Atomic force microscope images of (a) an as-prepared film (99% transmittance at 550 nm), and the film after spraying of (b) cyclohexane and (c) toluene. (d) Height profile of the nanotube bundles indicated by the inverted triangles in (a–c). The left and right images in (a), (b), and (c) are the height and phase images, respectively. Deformed SWCNT bundles are indicated by arrows in (a). The green dotted circles in (c) indicate embedded SWCNT bundles after deposition of toluene because of swelling of PET [13].

fundamental transverse electric (TE_{10}) mode $(E_z = 0)$ with a frequency of 2450 MHz, so the microwave electric field (E_y) is sinusoidally distributed along the x- and z-axes and constant along the y-axis. Immediate flash Ohmic heating with an energy conversion of greater than 99% can be realized because the microwave electric field is parallel to the overall SWCNT film and can efficiently induce a fast oscillating current in the film. The amplitude of the conduction current density, J_s, induced on the CNT film by the microwave electric field intensity, E_{MW}, may be described as follows [16]:

$$J_s = \sigma_{CNT} E_{MW},\tag{3}$$

where σ_{CNT} is the electric conductivity of the SWCNT film.

Figure 9a shows the surface temperature and R_s changes of the SWCNT film by varying the irradiation time. The surface temperature of the SWCNT film is dramatically increased after

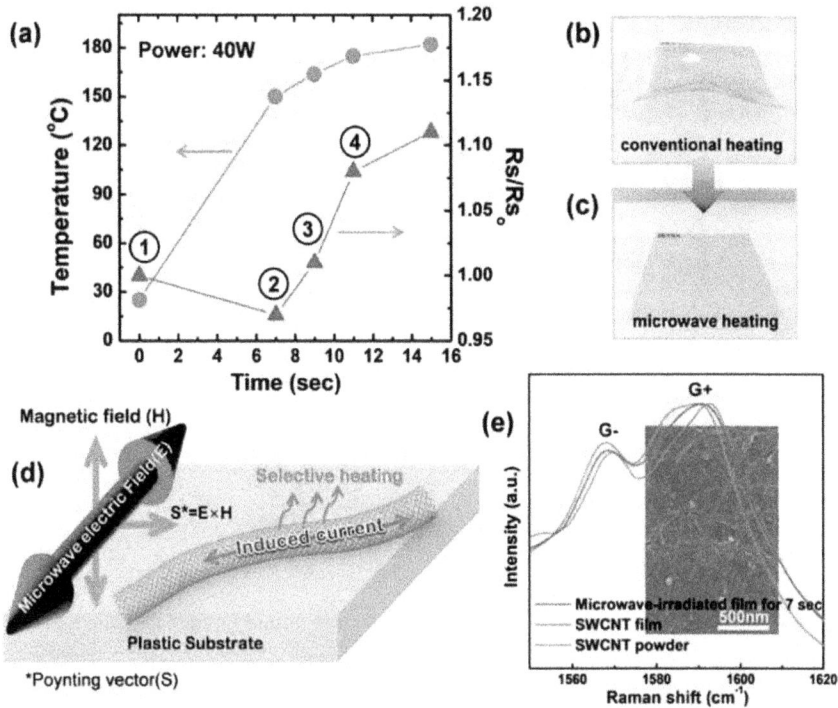

Figure 9. (a) Measured surface temperatures and R_s changes of SWCNT films on PC substrates as a function of microwave irradiation time. (b) The SWCNT film on PC heated in a conventional heating oven at 150°C. (c) The SWCNT film irradiated with microwaves. (d) Scheme of microwave-irradiated selective heating of CNTs on a plastic substrate, wherein a rapidly oscillating current induced along the CNTs is efficiently generated by the microwave electric field parallel to the SWCNT film. (e) Raman spectra of SWCNT powder and SWCNT films on PC before and after microwave irradiation for 7 s. Inset SEM image shows the microwave-nanowelded SWCNT film on the PC substrate [17].

7 s irradiation at 40 W without heat deflection. Of interest is that the R_s decreased after 7 s of irradiation, due to the occurrence of chemical welding. The Raman spectra in **Figure 9e** show the strain relaxation of the SWCNT network. The SEM image also shows clearly that the SWCNTs are welded or embedded in the plastic substrate. Importantly, the MW-irradiated SWCNT networks are protected by a self-passivation layer that protects the nanotubes from water molecules. The R_s values of the SWCNT films increase by less than 10% at 80°C and 90% relative humidity, despite embedding of the nanotubes in the plastic substrates.

3.3. CNT-induced migration of AgNW networks into plastic substrates

AgNW-based TCFs are not very environmentally stable without some form of passivation. If the AgNW network can be welded onto a thermoplastic substrate, it can be self-passivated, as was accomplished with SWCNT film. However, the surface tension of AgNWs (~500 mN/m of liquid silver in air) is much different from that of the hydrophobic PC substrate (~34.2 mN/m), which prevents the AgNWs from completely embedding in the plastic substrate, as illustrated in **Figure 10**. This surface tension mismatch can be solved by deposition of SWCNTs onto

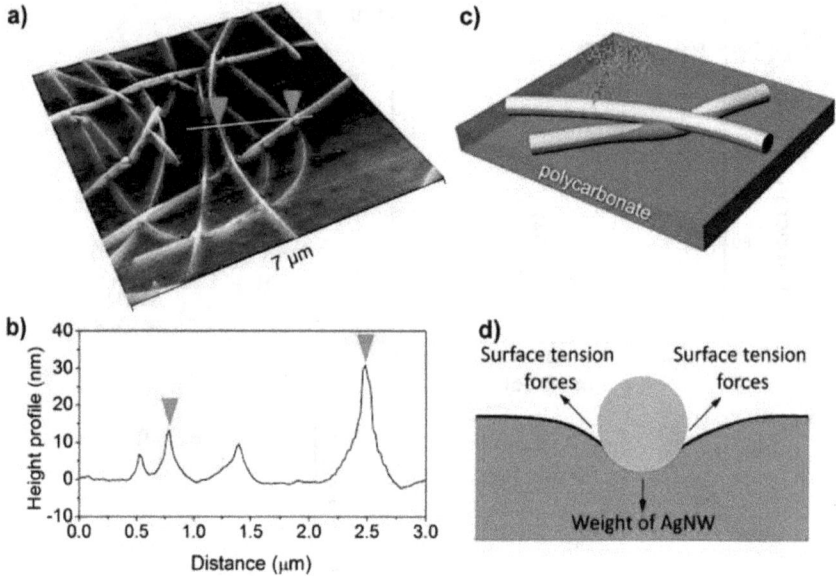

Figure 10. (a) AFM image and (b) height profile of the AgNW film after thermal treatment at 150°C for 3 h on a PC substrate. (c, d) Schematic illustration of the limited migration of AgNW networks into the plastic substrate due to a surface tension mismatch [18].

the AgNW network because of the low surface tension of CNTs (40–80 mN/m). Therefore, SWCNTs can trigger the migration of AgNWs into plastic substrates by thermal or chemical treatment. Moreover, the high thermal electrical conductivity of the SWCNT can promote the self-passivation of AgNWs by stable Joule heating of the film with an applied DC voltage. **Figure 11** shows the surface morphology of the SWCNT-overcoated AgNW film after a voltage of 20 V was applied. In stark contrast to AgNWs in AgNW film shown in **Figure 10a**, AgNWs were fully embedded in the plastic substrate by electrical heating. Atomic force microscopy (AFM) height profiles also demonstrate the embedding of the AgNW–SWCNT network in the plastic substrate. This self-passivation of AgNW networks assisted by SWCNTs with electrical heating improved the mechanical and hydrothermal stability of the film.

3.4. Interfacial engineering with GO for AgNW TCFs

In terms of the applications of metal nanowire networks, interfacial engineering is an important step to improve their performance with respect to electrical conductivity, environmental stability, surface roughness, and work function modulation. In particular, interfacial engineering of AgNW film can affect the opto-electrical performance because of junction formation in the network. In this study, HOGO nanosheets were utilized for efficient thermal joining of AgNW networks on thermoplastic substrates (**Figure 12a**). **Figure 12b** shows the R_s changes of the AgNW network films on bare PC, GO-modified PC, and glass after heating at 150°C with increasing exposure time. The R_s was dramatically reduced by thermal treatment via a junction joining of the networks. Importantly, the R_s decrease of the AgNW film was

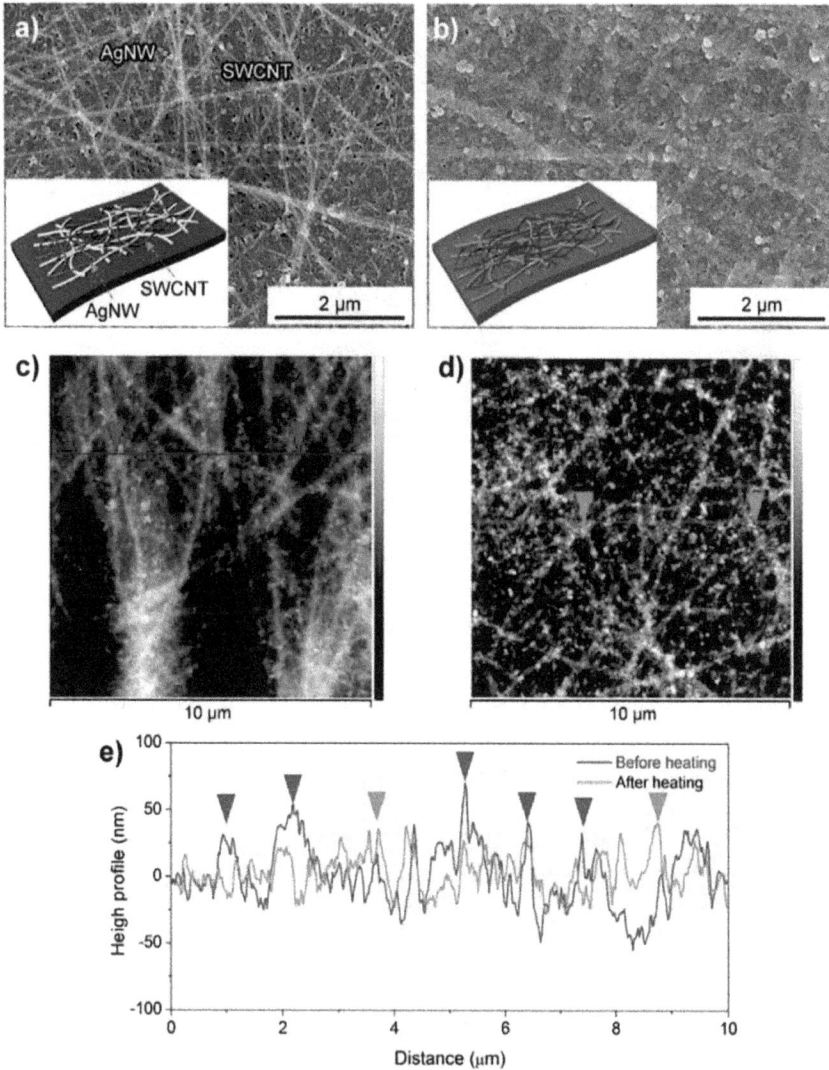

Figure 11. Field emission SEM images of AgNW overcoated with SWCNTs (a) before and (b) after heating under a current flow of thin film heater. AFM images of the same film (c) before heating and (d) after heating. (e) Height profile of the SWCNT-overcoated AgNW film under a current flow [18].

more efficient on GO-modified PC than on bare PC and glass. Interestingly, the changed R_s of AgNW films on PC was stable even after heating for 180 min, while the R_s of the AgNW film on glass gradually increased, even after 30 min, due to air oxidation. This result provides an opportunity to obtain high-performance AgNW TCFs by a combination of thermal welding and junction joining of AgNW networks. SEM and AFM images in **Figure 12** show clearly that on GO nanosheets, limited embedding or welding of AgNWs was observed. This demonstrates the more efficient reduction of R_s of AgNWs on the GO-modified PC.

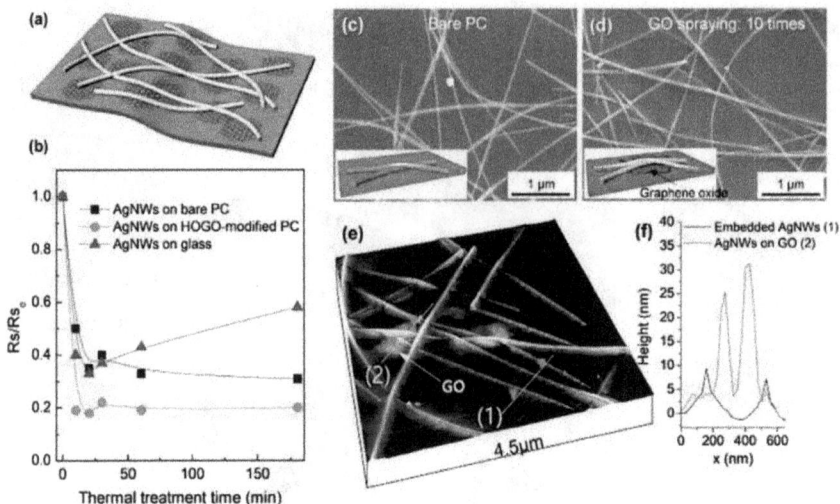

Figure 12. (a) Scheme showing AgNW film on GO-modified PC. (b) R_s changes of AgNW films on bare PC and GO-modified PC, and on glass after heating at 150°C by varying the exposure time. (c, d) SEM images of AgNW films on (c) bare PC and (d) GO-modified PC substrates after heating at 150°C for 1 h. (e) AFM image of AgNW networks on GO-modified PC. (f) Height profiles of embedded AgNWs and AgNWs floated on the GO nanosheet indicated in (e) as numbers [19].

4. High-performance TCFs by hybridization of 1D or 2D materials

4.1. Graphene oxide-modified SWCNT-based TCFs

SWCNT-based TCFs with a low haze value are suitable for highly transparent opto-electronic devices. However, for achievement of a low R_s value of the films, one challenge is the development of an efficient and stable dopant. In addition, their high porosity and hydrophobic surface properties are a drawback as an electrode material in opto-electronic devices. In this context, we introduced easily deformable GO nanosheets containing electron-withdrawing groups on the basal plane and edges, which can give a p-type doping effect on the SWCNT film. **Figure 13** shows that the R_s of the SWCNT film can be dramatically reduced by up to 40% compared to the as-prepared SWCNT film by deposition of GO solution onto the film by spraying. The efficiency of R_s reduction depends on the lateral sizes of the GO nanosheets. Small-sized GO nanosheets prepared by decanting the first supernatant (S1) by centrifugation were more efficient than larger GO nanosheets. As shown in **Figure 14**, the SWCNT bundles are easily wrapped with small GO nanosheets, while larger GO nanosheets can be freestanding between SWCNT networks. This means that densification of the SWCNT network is more efficient using small GO than large GO. The reduction of porosity and junction resistance of the SWCNT network can have a positive effect on the decrease of R_s. Moreover, the effect of p-type doping by GO is clearly shown in Raman spectra (**Figure 14c** and **d**). An upshift of 3.5 cm^{-1} in the G+ band for the semiconducting SWCNTs by small GO nanosheets (S1) demonstrates p-type doping of the SWCNTs from the GO nanosheets via a charge transfer mechanism.

To evaluate the opto-electrical performance of a GO-SWCNT electrode on PET, organic photovoltaic (OPV) cells with a PET/GO-SWCNT/PEDOT-PSS/active layer/LiF/Al structure were

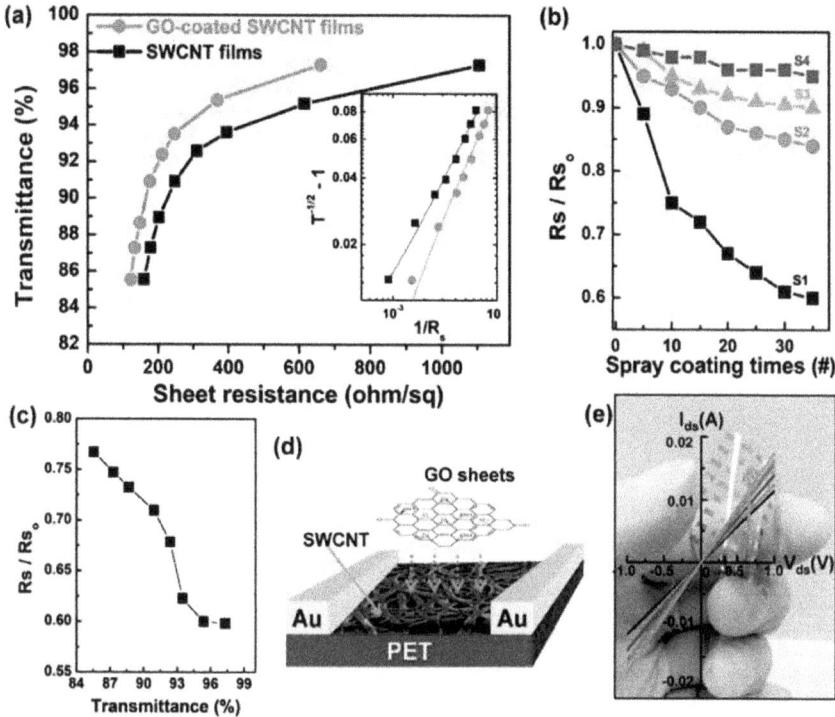

Figure 13. (a) R_s versus transmittance plots of SWCNT films before and after deposition of GO nanosheets. (b) Relative R_s changes of SWCNT films by increasing the number of spray coatings of GO solution obtained by centrifugation (the first to fourth supernatant solutions are denoted as S1 to S4). (c) Relative R_s as a function of the SWCNT film transmittance showing thickness dependence of GO deposition on R_s changes of the film due to contact area change between GO and the SWCNT bundle. (d) The I-V measurement scheme performed on SWCNT films after deposition of the GO solution. (e) Photo image of a gold-patterned SWCNT film and I-V plots for SWCNT films by increasing the amount of deposited GO solution (in the direction of the arrow) [20].

fabricated (**Figure 15**). For fabrication of the layered structure of the OPV cells, the wettability of the electrode on the upper loaded aqueous PEDOT:PSS solution is important. As shown in **Figure 15a**, the hydrophobic SWCNT film was converted to hydrophilic by deposition of hydrophilic GO nanosheets. Moreover, importantly, the work function of the SWCNT film changed from 4.7 to 5.05 eV by deposition of S1-GO nanosheets, which induces a facile hole injection from the HOMO of P3HT (5.0 eV) to the electrode. The resultant device performance with the GO-modified SWCNT anodes shows a significant enhancement in overall photovoltaic performance compared to devices fabricated on pristine SWCNT electrodes, as shown in **Figure 15d**.

4.2. Electrically stable SWCNT/AgNW hybrid TCFs

Under high current flow, metal NW scan be disrupted by Joule heating at the junction due to a relatively high junction resistance between metal NWs. Self-joining of NW network junctions can solve this problem via post-treatment. Another approach is to interconnect the NWs with other conducting materials or metal oxides. For more efficient processing of metal NW-based TCFs, we need to exclude additional steps, such as irradiation with light, heating at high

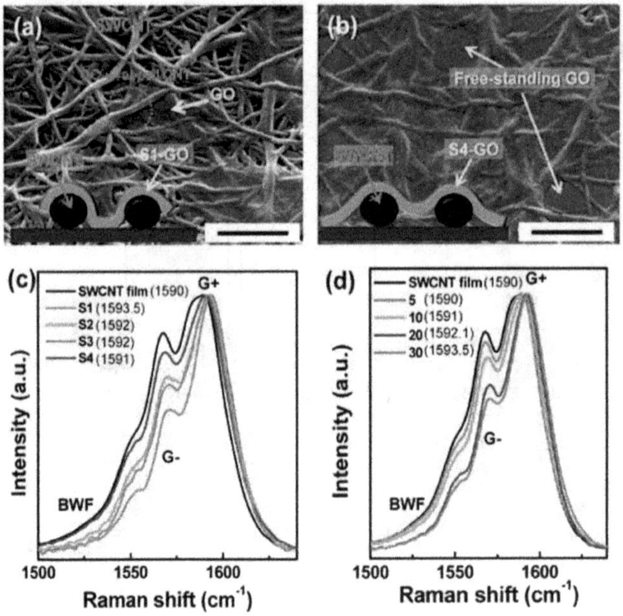

Figure 14. Tilted SEM images of SWCNT surfaces coated with (a) S1-GO nanosheets and (b) S4-GO nanosheets. Inset schemes show the structure of the GO-coated SWCNT networks. (c) Raman spectra of a pristine SWCNT film and films coated with S1, S2, S3, and S4 using a spray-coater 20 times. (d) Raman spectra of SWCNT films coated with S1-GO by increasing the number of coating layers from 5 to 30. Values in brackets in (c) and (d) indicate G+ band position. Scale bars in (a) and (b) are 300 nm [20].

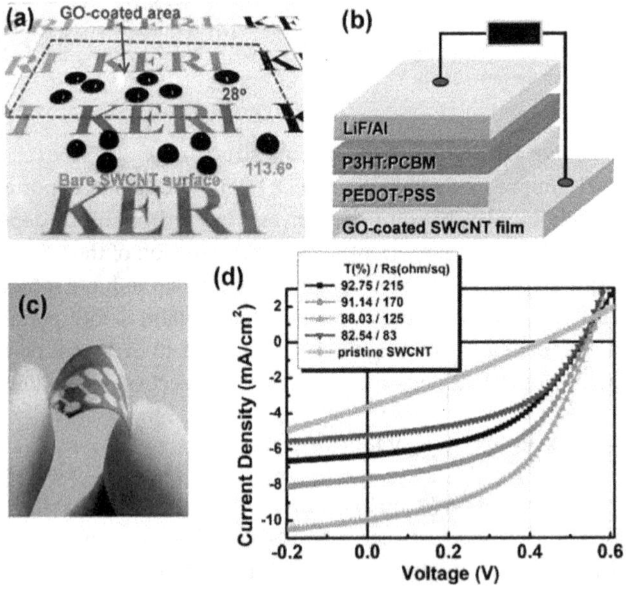

Figure 15. (a) PEDOT:PSS solution drop images on a, b are SWCNT surface and on the GO-coated area (dotted area). (b) Schematic structure and (c) photo image of OPV cell. (d) Current density (J) versus voltage (V) characteristics of pristine SWCNTs and GO-modified SWCNT photovoltaic cells under 100 mW/cm^2 AM 1.5G spectral illumination at various transmittance and R_s values [20].

temperatures, and the removal of surfactant molecules after the deposition of AgNWs or AgNW hybrid materials. Thus, we suggest that a small amount of SWCNTs can stabilize the AgNW networks under current flow without post-treatment. To realize this, the major challenge is the fabrication of a stable dispersion of SWCNTs in liquid medium without dispersant molecules that can be removed after deposition. To solve this issue, the SWCNTs were functionalized with quadruple hydrogen bonding (QHB) motifs of 2-ureido-4[1H]pyrimidinone (UHP) moieties through a previously reported sequential coupling reaction [21]. The AgNW/SWCNT mixture solution was easily prepared by direct mixing of the aqueous AgNW solution with a paste of SWCNTs functionalized with UHP (UHP-SWCNTs) by shaking, as shown in **Figure 16a**. The spray-coated AgNW/SWCNT hybrid film has an R_s value of ~20 Ω/sq. and $T > 90\%$ and was used to fabricate transparent film heaters to investigate the effect of SWCNTs on the electrical stability of the AgNW films under current flow. Notably, the breaking up of AgNWs at junctions was observed at 9 V (**Figure 17a**), which might have been induced by rapid joule heating at the junctions because of the high junction resistance of the AgNWs ($R_{11} \approx 10^3$–10^9 Ω). In stark contrast, after hybridization with SWCNTs, a new current pathway through the AgNW-SWCNT junction may be formed because of the relatively low contact resistance between the AgNW and SWCNT ($R_{12} \approx 10^3$ Ω) when compared to R_{11}, resulting in the formation of stable network films even at 15 V. Moreover, a very small work function difference between AgNW and UHP-SWCNTs, based on the Φ values of AgNW (4.1 eV) and UHP-SWCNTs (4.3 eV), can promote the current pathway through the AgNW-SWCNT junction (**Figure 18**).

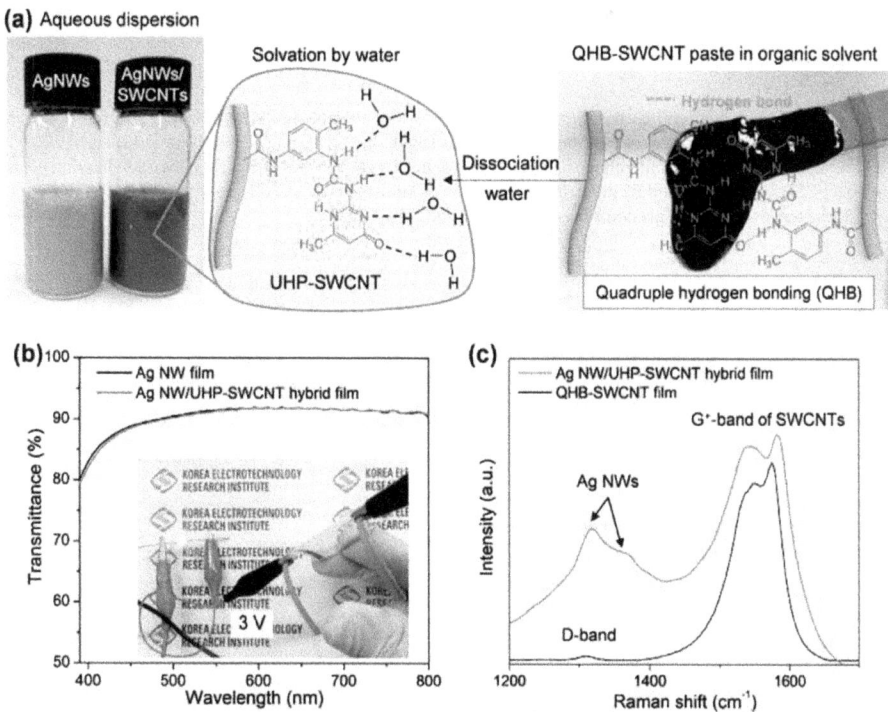

Figure 16. (a) Preparation of AgNW/SWCNT solution by direct mixing of aqueous AgNW solution and UHP-functionalized SWCNTs. (b) Optical transmission of the AgNW and AgNW/SWCNT hybrid films with $R_s \approx 20$ ohm/sq. fabricated by spraying. Inset image shows the lighting of an LED lamp at 3 V on bendable AgNW/SWCNT hybrid film on a polycarbonate substrate. (c) Raman spectra of the QHB-SWCNT film prepared by paste and AgNW/UHP-SWCNT hybrid films fabricated by mixture inks [22].

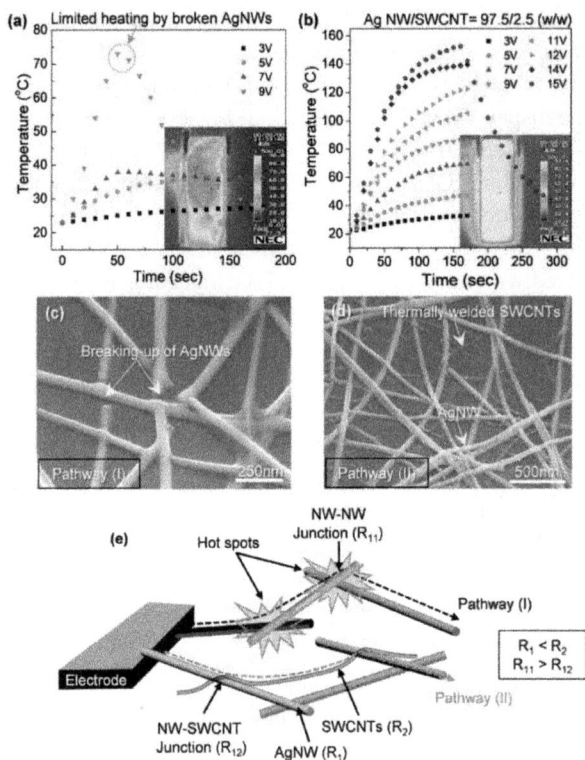

Figure 17. (a, b) Time-dependent temperature profiles of (a) AgNW and (b) AgNW/SWCNT hybrid films. The inset images are infrared thermal images of the film heaters. (c, d) Tilted SEM images of (c) AgNW and (d) AgNW/SWCNT hybrid films after heating at an input voltage of 9 V. (e) Schematic of AgNW/SWCNT hybrid networks showing possible current flow pathways (I, II). R1 and R2 indicate the resistivity of AgNWs and SWCNTs, respectively. R11 or R12 indicate the contact resistances between AgNWs or between AgNW and SWCNTs.

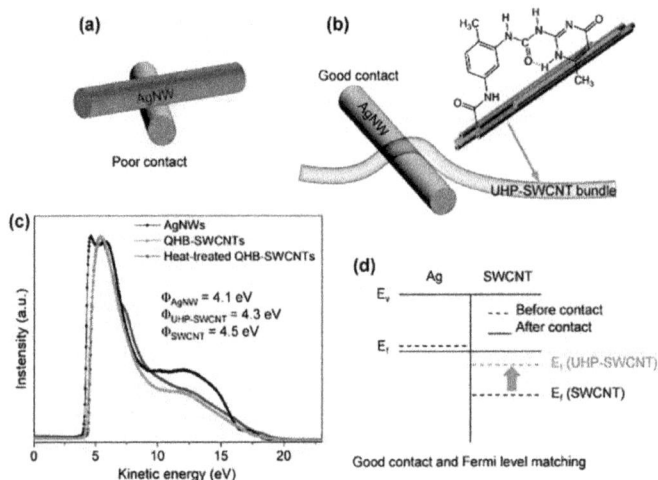

Figure 18. (a, b) Schematic diagram showing poor contact between AgNWs (a) and good contact between AgNW and UHP-SWCNTs. (c) Ultraviolet photoelectron spectroscopy spectra of AgNW, UHP-SWCNT, and thermally treated UHP-SWCNT films. (d) Schematic showing the reason for the current pathway through SWCNTs in terms of the work function.

5. Summary

We have briefly reviewed recent research progress on TCF technologies based on SWCNTs, AgNWs, and GO nanosheets via interfacial engineering and hybridization strategies. One-dimensional (1D) conducting nanomaterials such as CNTs and metal nanowires have been studied intensively because of their fascinating properties and offer tremendous potential for flexible opto-electronic applications in touch screen panels, flexible displays, solar cells, thin film heaters, signage, etc. To realize these applications, we need to develop high-performance TCFs with flexibility using a low-temperature process with scalable processing techniques on flexible plastic substrates. In this chapter, therefore, a scalable spray coating process using SWCNTs and AgNW solutions was introduced by demonstrating the wettability of the solution on surface energy-controlled substrates. One of the most important strategies for high-performance TCFs is interfacial engineering. Matching the interfacial tension between top-coating materials and the film is an important practical concept for fabrication of passivated TCFs that are environmentally stable at high humidity and temperature, as well as to improve their opto-electrical properties. Moreover, rational use of GO nanosheets and SWCNTs can improve AgNW network TCFs by welding in plastic substrates and efficient junction joining of AgNW junctions. Chemical or thermal welding of SWCNT networks is also useful for self-passivation of films on thermoplastic substrates.

In addition, recently developed AgNW/SWCNT hybrid TCF technologies can be commercially used to fabricate large area flexible TCFs by a roll-to-roll process because of fabrication of coating solutions without additional dispersant molecules.

For large opto-electronic devices with flexibility and stretchability, there are still many challenging issues for commercial application, including newly designed anisotropic conducting materials and their solution processing.

Acknowledgements

This work was supported by the Center for Advanced Soft-Electronics as Global Frontier Project (2014M3A6A5060953) funded by the Ministry of Science, ICT and Future Planning and by the Primary Research Program (18-12-N0101-18) of the Korea Electrotechnology Research Institute.

Author details

Joong Tark Han* and Geon-Woong Lee

*Address all correspondence to: jthan@keri.re.kr

Korea Electrotechnology Research Institute, Republic of Korea

References

[1] Ye S, Rathmell AR, Chen Z, Stewart IE, Wiley BJ. Metal nanowire networks: The next generation of transparent conductors. Advanced Materials. 2014;**26**:6670-6687. DOI: 10.1002/adma.201402710

[2] Kim Y, Ryu TI, Ok K-H, Kwak M-G, Park S, Park N-G, Han CJ, Kim BS, Ko MJ, Son HJ, Kim J-W. Inverted layer-by-layer fabrication of an ultraflexible and transparent Ag nanowire/conductive polymer composite electrode for use in high-performance organic solar cell. Advanced Functional Materials. 2015;**25**:4580-4589. DOI: 10.1002/adfm.201501046

[3] Xu F, Wu M-Y, Safron NS, Roy SS, Jacobberger RM, Bindl DJ, Seo J-H, Chang T-H, Ma Z, Arnold MS. Highly stretchable carbon nanotube transistors with ion gel gate dielectrics. Nano Letters. 2014;**14**:682-686. DOI: 10.1021/nl403941a

[4] Jeong I, Chiba T, Delacou C, Guo Y, Kaskela A, Reynaud O, Kauppinen EI, Maruyama S, Matsuo Y. Single-walled carbon nanotube film as electrode in indium-free planar heterojunction perovskite solar cells: Investigation of electron-blocking layers and dopants. Nano Letters. 2015;**15**:6665-6671. DOI: 10.1021/acs.nanolett.5b02490

[5] Li Z, Boix PP, Xing G, Fu K, Kulkarni SA, Batabyal SK, Xu W, Cao A, Sum TC, Mathews N, Wong LH. Carbon nanotubes as an efficient hole collector for high voltage methylammonium lead bromide perovskite solar cells. Nanoscale. 2016;**8**:6352-6360. DOI: 10.1039/C5NR06177F

[6] Woo JS, Lee G-W, Park S-Y, Han JT. Realization of transparent conducting networks with high uniformity by spray deposition on flexible substrates. Thin Solid Films. 2017;**638**: 367-370. DOI: 10.1016/j.tsf.2017.08.010

[7] Hu L, Hecht DS, Grüner G. Percolation in transparent and conducting carbon nanotube networks. Nano Letters. 2004;**4**:2513-2517. DOI: 10.1021/nl048435y

[8] Lyons PE, De S, Blighe F, Nicolosi V, Pereira LFC, Ferreira MS, Coleman JN. The relationship between network morphology and conductivity in nanotube films. Journal of Applied Physics. 2008;**104**:044302/1-0044302/8. DOI: 10.1063/1.2968437

[9] Hecht D, Hu LB, Grüner G. Conductivity scaling with bundle length and diameter in single walled carbon nanotube networks. Applied Physics Letters. 2006;**89**:13112/1-13112/3. DOI: 10.1063/1.2356999

[10] Simien D, Fagan JA, Luo W, Douglas JF, Migler K, Obrzut J. Influence of nanotube length on the optical and conductivity properties of thin single-wall carbon nanotube networks. ACS Nano. 2008;**2**:1879-1884. DOI: 10.1021/nn800376x

[11] Nirmalraj PN, Lyons PE, De S, Coleman JN, Boland JJ. Electrical connectivity in single-walled carbon nanotube networks. Nano Letters. 2009;**9**:3890-3895

[12] Han JT, Kim JS, Jeong HD, Jeong HJ, Jeong SY, Lee G-W. Modulating conductivity, environmental stability of transparent conducting nanotube films on flexible substrates by interfacial engineering. ACS Nano. 2010;**4**:4551-4558

[13] Han JT, Kim JS, Lee SG, Bong H, Jeong HJ, Jeong SY, Cho K, Lee G-W. Chemical strain-relaxation of single-walled carbon nanotubes on plastic substrates for enhanced conductivity. Journal of Physical Chemistry C. 2011;**115**:22251-22256. DOI: 10.1021/nl9020914

[14] Imholt TJ, Dyke CA, Hasslacher B, Perez JM, Price DW, Roberts JA, Scott JB, Wadhawan A, Ye Z, Tour JM. Nanotubes in microwave fields: Light emission, intense heat, outgassing, and reconstruction. Chemistry of Materials. 2003;**15**:3969-3970. DOI: 10.1021/cm034530g

[15] Ye Z, Deering WD, Krokhin A, Roberts JA. Microwave absorption by an array of carbon nanotubes: A phenomenological model. Physical Review B. 2006;**74**:075425. DOI: 10.1103/PhysRevB.74.075425

[16] Metaxas AC, Meredith RJ. Industrial Microwave Heating. London: Peter Peregrinus Ltd.; 1988. ISBN-0-906048-89-3

[17] Han JT, Kim D, Kim JS, Seol SK, Jeong SY, Jeong HJ, Chang WS, Lee G-W, Jung S. Self-passivation of transparent single-walled carbon nanotube films on plastic substrates by microwave-induced rapid nanowelding. Applied Physics Letters. 2012;**100**:163120/1-163120/4. DOI: 10.1063/1.4704666

[18] Woo JS, Kim BK, Kim HY, Lee G-W, Park SY, Han JT. Carbon nanotube-induced migration of silver nanowire networks into plastic substrates via Joule heating for high stability. RSC Advances. 2016;**6**:86395-86400. DOI: 10.1039/C6RA17771A

[19] Woo JS, Sin DH, Kim H, Jang JI, Kim HY, Lee G-W, Cho K, Park S-Y, Han JT. Enhanced transparent conducting networks on plastic substrates modified with highly oxidized graphene oxide nanosheets. Nanoscale. 2016;**8**:6693-6699. DOI: 10.1039/C5NR08687F

[20] Han JT, Kim JS, Jo SB, Kim SH, Kim JS, Kang B, Jeong HJ, Jeong SY, Cho K, Lee G-W. Graphene oxide as a multi-functional p-dopant of transparent single-walled carbon nanotube films for optoelectronic devices. Nanoscale. 2012;**4**:7735-7742. DOI: 10.1039/C2NR31923C

[21] Han JT, Jeong BH, Seo SH, Roh KC, Kim S, Choi S, Woo JS, Kim HY, Jang JI, Shin D-C, Jeong S, Jeong HJ, Jeong SY, Lee G-W. Dispersant-free conducting pastes for flexible and printed nanocarbon electrodes. Nature Communications. 2013;**4**:2491. DOI: 10.1038/ncomms3491

[22] Woo JS, Han JT, Jung S, Jang JI, Kim HY, Jeong HJ, Jeong SY, Baeg K-J, Lee G-W. Electrically robust metal nanowire network formation by in-situ interconnection with single-walled carbon nanotubes. Scientific Reports. 2014;**4**:4804. DOI: 10.1038/srep04804

A Facile Fabrication Criteria of Carbon Nanotube for Transparent Conducting Films Application

Chapter 3

Transparent Conducting Thin Film Preparation of Carbon Nanotube

Xiaogang Sun, Jie Wang, Wei Chen, Xu Li,
Manyuan Cai, Long Chen, Zhiwen Qiu,
Yapan Huang, Chengcheng Wei, Hao Hu and
Guodong Liang

Additional information is available at the end of the chapter

http://dx.doi.org/10.5772/intechopen.79164

Abstract

Transparent conducting films have a wide range of applications in the fields of flat panel displays, solar cells, and touch panels for their both good conductivity and light transmittance. Carbon nanotubes (CNTs) transparent conducting film has become a potential alternative for next-generation transparent conducting film systems owing to high conductivity, light transmittance and flexibility. The multiwalled carbon nanotubes (MWCNTs) conductive liquid was prepared by dispersing MWCNTs in alcohol through ultrasonic and high-speed shearing process with an addition of carbon nanotube alcohol dispersant (TNADIS) as the dispersant. The transparent conducting film was fabricated on polyethylene terephthalate (PET) transparent film by spin-coating process. The film was used as interlayer between the electrode and the separator to improve electrochemical performance of lithium-sulfur (Li-S) batteries.

Keywords: multiwalled carbon nanotubes, transparent conducting film, lithium-sulfur batteries, transmittance

1. Introduction

Carbon nanotubes have been the hotpot of scientific research ever since their discovery. Due to their unique structure, carbon nanotubes (CNTs) have shown outstanding performance in electromagnetics, mechanics, heat and optics [1–5], which have made them attractive in lithium ion batteries, supercapacitors, composite materials and many other aspects [2–6]. At present, carbon nanotubes have been produced in large scale. However, the carbon nanotubes entangled with each other and shown severe agglomeration effect [7, 8]. The carbon nanotubes are nanoscale materials and the specific surface area is large, the surface energy is high, and there is a great Van der Waals force between the carbon nanotubes [9–15]. In addition, the carbon nanotubes exhibited a structure of one-dimensional tubular and the aspect ratio is relatively large, they have similar interlocking characteristics to fibers, which result in easy agglomeration of carbon nanotubes. In order to solve the technical problem and obtain a stable carbon nanotube dispersion, many dispersion methods are introduced to prepare the dispersion liquid of the carbon nanotube. Physical dispersion methods include grinding, ball milling, ultrasonic, and high-speed shear. The chemical dispersion methods include strong acid and alkali treatment and the addition of dispersant [16–18]. Each dispersion method has its own advantages but all have some drawbacks that make it difficult to produce a very stable dispersion of carbon nanotubes.

Transparent conductive films are widely used in the fields of flat panel displays, solar cells, touch panels owing to good electrical conductivity and light transmission. Currently, many transparent conductive films are studied: metal film, n-type transparent conductive oxide film, p-type transparent conductive oxide film, special film system (TiN conductive film, etc.) and multilayer film system. Carbon nanotubes (CNTs) have also become the focus of research due to their good properties in conductivity, light transmission and flexibility. Therefore, the CNTs transparent conducting films have also become the focus of research. The dispersibility of carbon nanotubes has an important influence on the quality of the film of the conductivity as well as transparency. In this chapter, MWCNTs ethanol conductive liquid was prepared by ultrasonic vibration and high-speed shearing process. The MWCNTs transparent conductive film was prepared by spin-coating.

The high theoretical capacity of 1675 mAh/g and high energy density of 2600 Wh/kg, lithium–sulfur (Li-S) batteries have become the most promising alternatives for next-generation electrochemical energy storage systems [19–22]. In addition, abundant resources, low cost and ecofriendliness of sulfur make Li-S batteries have higher commercial value [23–25]. However, the actual capacity of the current lithium-sulfur battery is greatly lower than the theoretical capacity and the cycle life is poor, which seriously hampered the practical application of lithium-sulfur battery [26]. The main reason is that the diffusion and dissolution of intermediate lithium polysulfides during cycling (Li_2S_n, $4 \leq n \leq 8$) which led to notorious shuttle effects. This resulted in high self-discharge, active material loss and low Coulombic efficiency [27–31]. In this work, we report multiwalled carbon nanotubes paper (MWCNTsP) as a current collector, MWCNTs transparent conductive film was used as the interlayer between the positive electrode and the separator [32, 33]. The new structure of Li-S battery retarded the dissolution and dispersion of lithium polysulfides (LPSs). The Li-S batteries with MWCNTs transparent conductive film showed high discharge capacity, excellent cycle stability and high sulfur loading.

2. Fabrication and characterization of MWCNTs

MWCNTs were synthesized by chemical vapor deposition with benzene being used as carbon feedstock, ferrocene as a catalyst precursor, thiophene as growth promotion agent, and

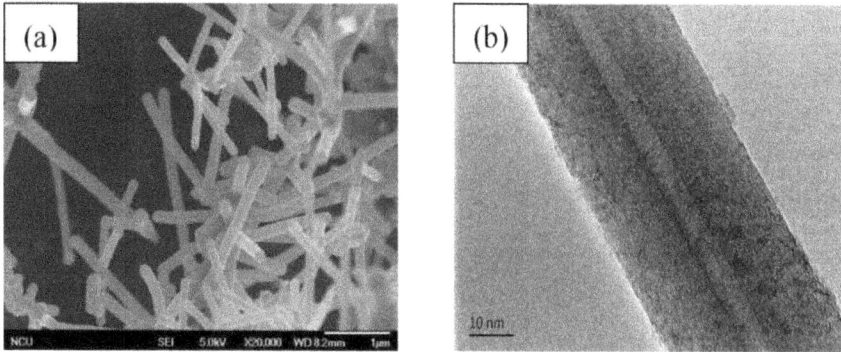

Figure 1. (a) SEM and (b) TEM images of MWCNT.

hydrogen as carrier gas. The reaction temperature was around 1200°C. The raw MWCNTs were treated at a high temperature of 3000°C for graphitization.

The SEM image of multiwalled carbon nanotube was shown in **Figure 1(a)**, it was observed that the MWCNTs have a one-dimensional tubular structure and are not entangled with each other. This suggested that the line MWCNTs can be easily dispersed in various matrix. The MWCNTs have also excellent mechanical and physicochemical properties. The TEM (**Figure 1(b)**) image shows that MWCNTs have one-dimensional tubular structure and the carbon atoms are arranged in a regular and orderly manner.

3. MWCNTs transparent conducting film

3.1. Experiment

3.1.1. Preparation of MWCNTs conductive liquid

The raw multiwalled carbon nanotubes (R-MWCNTs) and graphitized multiwalled carbon nanotubes (G-MWCNTs) were, respectively, ball-milled in a ball mill for 2 h. The TNADIS dispersant was dissolved in anhydrous ethanol and its mass concentration was 0, 0.025, 0.05, 0.1, 0.2, 0.4, 0.6, 0.8, and 1%, respectively. Then, two types of MWCNTs were, respectively, added to the above solution with a concentration of 1 wt.%. The dispersion liquid of MWCNTs was prepared by ultrasonically dispersing for 30 min and high-speed shearing for 1 h. The R-MWCHTs are marked as 1#-8# and the G-MWCNTs are marked as 9–16#.

3.1.2. Performance testing

Raman spectroscopic analysis, transmission electron microscopy (TEM) and scanning electron microscopy (SEM) were used to analyze and observe the morphology and structure of MWCNTs. The optimal addition ratio of dispersant was determined by measuring the precipitation after centrifugation. In addition, the stability of the conductive liquid was observed and analyzed. The stability of the conductive liquid of MWCNTs was characterized by detecting the Tyndall effect after the conductive liquid was left standing for 5 months.

3.2. Results and discussion

3.2.1. Raman spectroscopy and TEM, SEM analysis

The Raman spectra of MWCNTs was shown in **Figure 2**, it can be used to analyze the crystallinity of MWCNTs. The relative intensity of D and G peaks (IG/ID) can reflect the degree of crystallization of MWCNTs samples. G-MWCNTs have a much higher IG/ID of 4.16 than R-MWCNTs (IG/ID = 0.67). The G peak at 1585 cm^{-1} is called the tangential stretching mode of the MWCNTs, which is a reflection of the degree of order, and similar peaks are observed in the graphite. The D peak at 1334 cm^{-1} is a reflection of the defect and disorder degree in the MWCNTs, and the D peak of the MWCNTs originates from the structural defects of the carbon nanotubes. Both the D peak after graphitization and the 2D peak at 2656 cm^{-1} were observed due to the double resonance process of the two resonant electron states of the MWCNTs.

The TEM image of raw MWCNTs was shown in **Figure 3(a)**, and it was observed that the arrangement of carbon atoms on the surface of R-MWCNTs is disordered. It indicated there are a lot of amorphous carbons and defects on surface of R-MWCNTs. After MWCNTs were graphitized at 3000°C, the carbon atoms shown a regular and orderly manner (**Figure 3(b)**).The G-MWCNTs have a high degree of crystallinity which is consistent with the results obtained by Raman spectroscopy. **Figure 2(c)** and **(d)** shows the SEM images before and after graphitization of MWCNTs, respectively. It can be seen that the MWCNTs used in the experiment are linear whisker carbon nanotubes. However, G-MWCNTs and R-MWCNTs are agglomerated before being dispersed due to van der Waals forces. From the SEM images, it can be seen that the graphitized carbon nanotubes have less impurities and higher purity.

3.2.2. Effect of dispersant TNADIS content on dispersion of MWCNTs

The precipitation mass versus dispersant content was exhibited (**Figure 4**), which depicts a linear reduce with increasing dispersant content until the minimum precipitation mass is achieved. Hereafter the sediment increased with increasing dispersant. The minimum

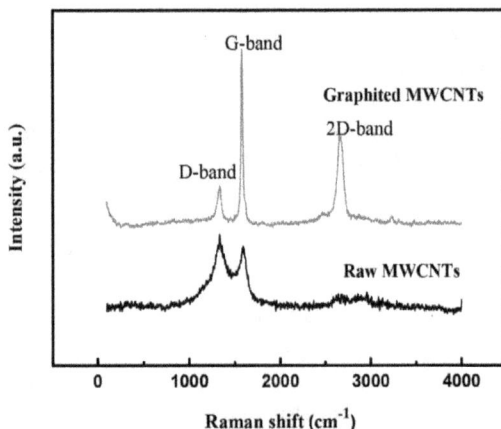

Figure 2. The Raman spectra of graphitization and raw MWCNTs.

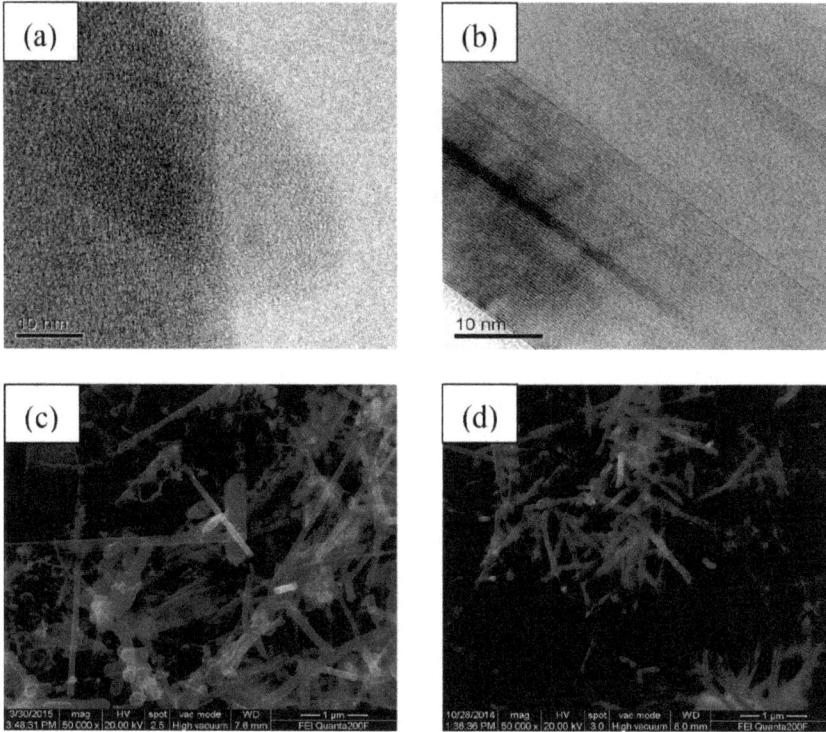

Figure 3. The TEM of R-MWCNTs (a), G-MWCNTs (b), the SEM of R-MWCNTs (c), and G-MWCNTs (d).

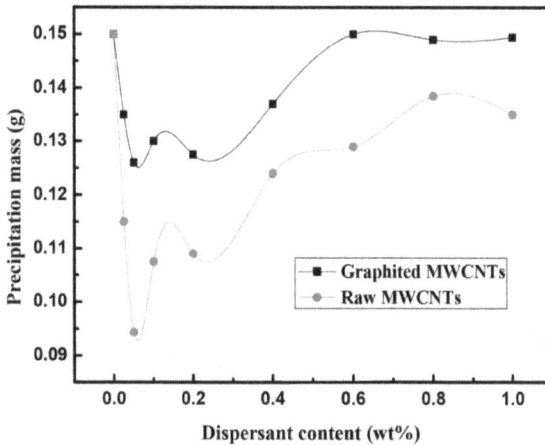

Figure 4. The precipitation mass of MWCNTs conducting liquids after centrifuged 10 min.

precipitation quality is around 0.0943 g with a dispersant of 0.05 wt.%. Van der Waals attraction between MWCNTs causes MWCNTs flocculation resulting in the formation of the aggregates and precipitation. The centrifugation treatment accelerated the sediment. The dispersant can disperse the MWCNTs aggregates and form stable dispersion liquid of ethanol. When the dispersant content exceeded a critical value, the micelles of the dispersant are formed and resulted in the aggregation of MWCNTs and more precipitates.

In addition, the precipitation produced by G-MWCNTs conducting liquid after centrifugation is higher than that of R-MWCNTs conductive liquid as shown in **Figure 4**. This is owing to that R-MWCNTs absorbed the dispersant molecules on the surface which checked the aggregation of R-MWCNTs. However, graphitization eliminated the surface defects of G-MWCNTs, which reduces the adsorption of dispersant molecules on the surface. Therefore, the conductive liquid of G-MWCNTs is easier to precipitate than R-MWCNTs after centrifugation treatment.

3.2.3. Observation of MWCNTs conductive liquid

After 5 days of standing of R-MWCNTs dispersion, only the 1# shown obvious delamination and sediment phenomenon and other samples maintained unchanging (**Figure 5(a)**). **Figure 5(b)** shows the conductive liquid of G-MWCNTs. The 9#(0% dispersant), 13#(0.4%), 14#(0.6%), 15#(0.8%) and 16#(0.1%) demonstrated obvious delamination and sediment. The 10#, 11# and 12# hold unchanging. The results showed that the stability of conductive liquid of R-MWCNTs was better than that of G-MWCNTs. This is attributed that R-MWCNTs have a lot of surface defects and absorbed much functional groups as OH. In addition, the stability of dispersion with different dispersant TNADIS mass fractions was analyzed (9#, 13#, 14#, 15#, and 16#). The results demonstrated the aggregation and sediment were enhanced with increasing dispersant which surpass the critical concentration.

MWCNTs conductive liquid (3#) with 0.05 wt.% TNADIS showed excellent stability and no obvious sedimentation was observed after resting at room temperature for 5 months. When a beam of light passes through the colloid, a bright path in the colloid can be observed from the direction of the incident light. This phenomenon is called the Tyndall effect. The Tyndall effect of the conductive liquid of R-MWCNTs was examined before and after standing for 5 months. The optical path can be seen in the conductive liquid (**Figure 6**). It is shown that the conductive liquid of R-MWCNTs has a good stability and still remained colloidal properties after standing for 5 months.

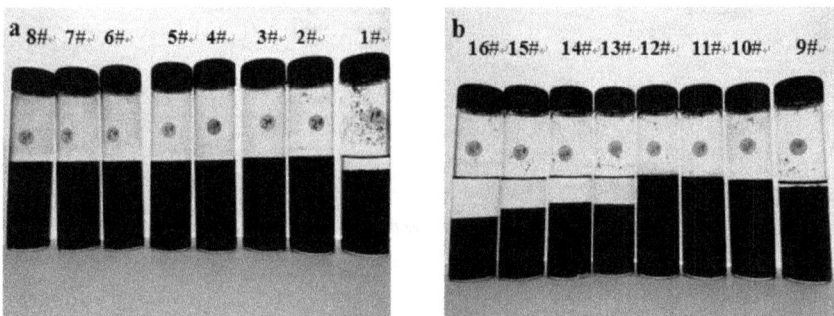

Figure 5. The raw (a) and graphitization (b) MWCNTs conducting liquid of stewing 5 days.

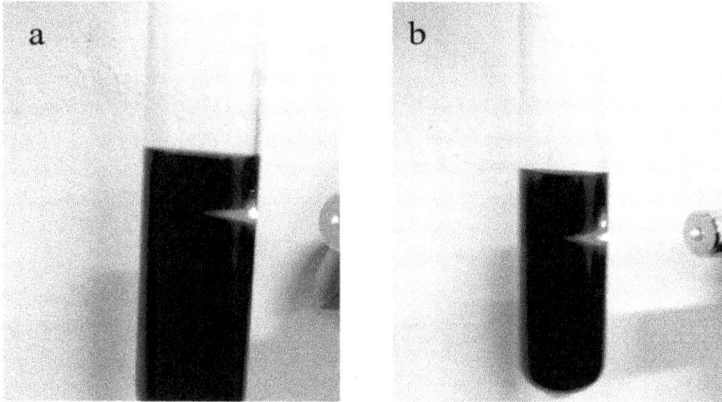

Figure 6. The Tyndall effect of R-MWCNTs conductive liquid before (a) and after (b) resting for 5 months.

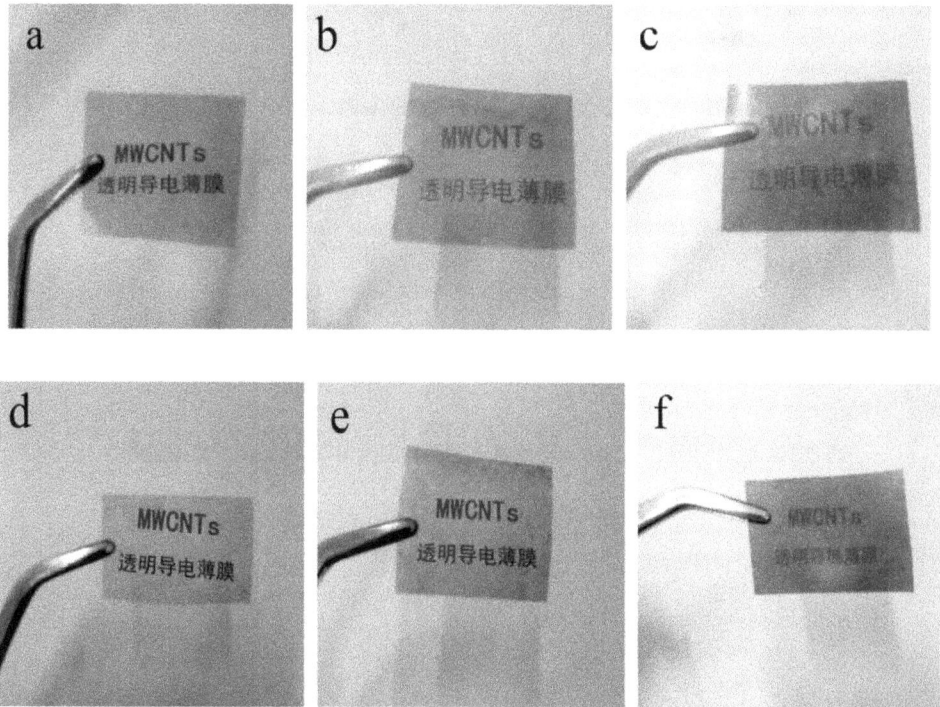

Figure 7. The R-MWCNTs (a, b, c) and G-MWCNTs (d, e, f) transparent conducting film.

3.2.4. Preparation and properties of MWCNTs transparent conductive film

The preparation method of MWCNTs transparent conductive films has high requirements on the dispersion properties of MWCNTs. The dispersion properties have a great influence on the film quality, electrical conductivity and transparency [34–41]. The 3#R-MWCNTs

The number of spin-coating		One	Twice	Three
R-MWCNTs	Square resistance kΩ/sq	103.3	10.6	3.7
	Transmittance (%)	68.3	57.9	52.8
G-MWCNTs	Square resistance kΩ/sq	53.6	2.8	0.34
	Transmittance (%)	68.9	58.1	53.3

Table 1. The square resistant and transmittance of MWCNTs transparent conducting film.

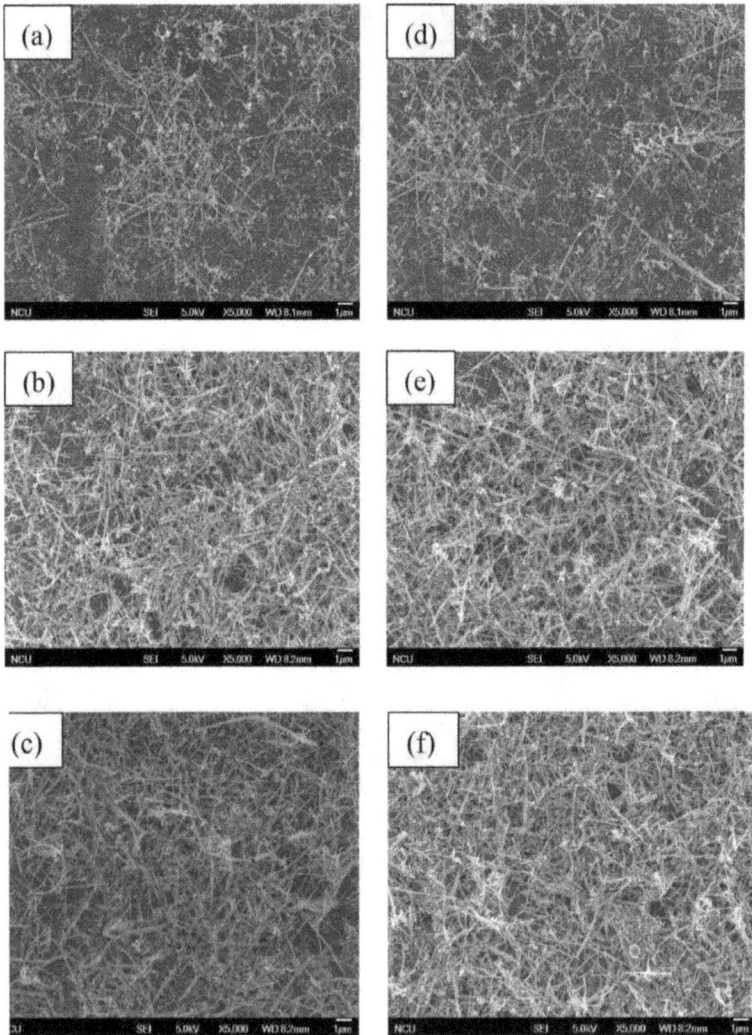

Figure 8. The SEM MWCNTs transparent conducting film. (a, b, c) R-MWCNTs spin-coating for once, twice, and thrice. (d, e, f) G-MWCNTs spin-coating for once, twice, and thrice.

conductive liquid and the 11#G-MWCNTs conductive liquid were applied onto the PET transparent film by spin-coating. The content of MWCNTs on the transparent film was controlled by controlling the number of spin-coating. After being spin-coated for once, twice, and thrice, respectively, the films were dried in vacuum drying oven. The square resistance of the dried MWCNTs transparent conductive film was measured with an ST-2258C multi-function digital four-probe tester.

Figure 7 shows spin-coated MWCNTs transparent conductive film. The transmittance of the film gradually decreases with increasing the number of spin-coating.

Compared with the resistance and transmittance of the R-MWCNTs and G-MWCNTs transparent conductive films, it was found that the transmittance of conductive films with the same number of spin-coating is almost the same. But, the conductivity of G-MWCNTs transparent conductive films was significantly better than that of R-MWCNTs as shown in **Table 1**. The main reason was that the G-MWCNTs obtained very high crystallinity and purity after high temperatures treatment of 3000°C. This resulted in higher electrical conductivity of G-MWCNTs transparent conductive films. In addition, as the number of spin-coating increases, the light transmittance of films gradually decreases and the conductivity increased.

Compared with the SEM image (**Figure 8**) of the conductive films, it was found that as the spin-on time increases, the MWCNTs gradually formed a continuous and dense mesh. The electronic transmission path was constructed, and the electrical conductivity was improved. After the spin-coating of G-MWCNTs conductive liquid for three times, the square resistance of the conductive film was 0.34 kΩ/sq.

4. MWCNTs transparent conducting film as interlayer for Li-S batteries

4.1. Experiment

4.1.1. Preparation of MWCNT paper collector

The MWCNT powder was dispersed in distilled water by sonication for 1 h and followed by high-speed shearing for 1 h with an addition of sodium dodecyl sulfate (SDS) as a surfactant. The cellulose fibers were prepared by smashing recycled papers in distilled water by high-speed shearing for 1 h. The MWCNT dispersion liquid and the cellulose fibers were mixed by high-shear emulsifier to form suspension for 2 h. The MWCNTs paper (MWCNTsP) was obtained by vacuum filtration through the suspension liquid of cellulose and MWCNTs. The MWCNTsP was rolled and tailored as current collector to host sulfur for cathodes.

4.1.2. Preparation of sulfur electrodes

Sulfur, MWCNTs and carbon black (CB) were mixed by balling for 1 h at 200 r/min. The slurry of sulfur was prepared by balling process with N-methyl-2-pyrrolidone (NMP) as solution and PVDF as binder. The ratio of S:MWCNT:CB:PVDF = 60:15:15:10. The blend slurry was coated on to porous MWCNTsP. Then the sulfur electrodes (S-MWCNTsP) were dried at 60°C under vacuum for 12 h.

The conductive liquid of MWCNTs was prepared by ultrasonication and high-shear process with NMP as solution and PVDF as binder. The ratio of MWCNTs:CB:PVDF = 60:30:10. Subsequently followed by overlaying the conductive liquid onto S-MWCNTsP electrodes, the obtained electrode with MWCNTs transparent conducting film (TCF@S-MWCNTsP).

4.1.3. Assembling of cell and electrochemical measurements

The tailored S-MWCNTsP and TCF@S-MWCNTsP electrodes were, respectively, used as working electrodes. Lithium foil was used as the counter electrode and Celgard 2300 was used as the separator. The solution of 1.0 M LiTFSI in DOL:DME (1,1, vol.) with 1.0%LiNO3 was utilized as the electrolyte. CR2025 coin-type cells were assembled in an Ar filled glove box (MBRAUN LABSTAR, Germany). The electrochemical characterization of the cells was measured by a cell tester (CT-4008-5V5mA-164). The galvanostatic charge-discharge current density was set at 0.2 to 5C. Electrochemical impedance spectroscopy (EIS) and cyclic voltammetry (CV) within a potential window of 1.6–2.8 V by an electrochemical workstation (CHI 660B) were measured.

4.2. Results and discussion

4.2.1. The SEM of MWCNTsP and TCF@S-MWCNTsP electrode

Top surface SEM of the MWCNTsP in **Figure 9(a)** displays its porous matrix and the coalescing fiber network. The MWCNTsP demonstrated homogenous incorporation of MWCNTs in the cellulose fiber network. Porous structure can effectively improve the carrying capacity of sulfur, and adsorbed PSs. The superior electrolyte absorbability and hierarchical open channel of MWCNTsP can store more sulfur and contributes to stabilize electrochemical reactions and anchored PSs within the MWCNTsP effectively and suppressing shuttle effects. The SEM image of TCF@S-MWCNTsP electrode in **Figure 9(b)** also shows that MWCNTs and S were

Figure 9. The SEM image of MWCNTsP (a) and TCF@S-MWCNTsP electrode (b).

dispersed evenly and well-connected, which should improve the conductivity of the electrodes and trapped more active material in its micropores toward the cathode side than that
toward the anode side. The excellent electrolyte immersion and active material encapsulation
also confirm the intimate connection between the insulating active material and the conductive matrix.

4.2.2. GTG of the TCF@S-MWCNTsP electrode

It can be indicated from **Figure 10** that the TCF@S-MWCNTsP electrode had a sulfur
content of around 14 wt.% according to the main weight loss at T1 interval, which was
attributed to sulfur sublimation. The weight loss at T2 interval which was attributed to
carbonization of paper fibers. The weight of the electrode is 26 mg and is the average
mass of the pole piece, and this can be calculated as the areal mass loading of sulfur in the
cathode is 2.4 mg/cm^2.

4.2.3. Cyclic voltammetric characteristics

Cyclic voltammetry (CV) plots of the TCF@S-MWCNTsP electrode for the initial three cycles
are shown in **Figure 11(a)**, recorded at a slow scan rate of 0.1 mV/s between 1.6 and 2.8 V. In
the first electrode scan, two characteristic reduction peaks at 2.29 and 1.99 V can be observed,
corresponding to the reduction of elemental sulfur (S_8) to long-chain PSs ($Li_2S_n, 4 \leq n \leq 8$)
and the subsequent formation of short-chain Li_2S_2/Li_2S, respectively. In the anodic sweep,
two oxidation peaks are observed at 2.43 and 2.47 V, which owing to the conversion of
short-chain to long-chain PSs, and the subsequent oxidization to S_8 [24, 42, 43].The third-
cycle CV curve is highly similar to the second-cycle curve, thus it can be expected that the
TCF@S-MWCNTsP electrode will show favorable cycling stability and high reversibility

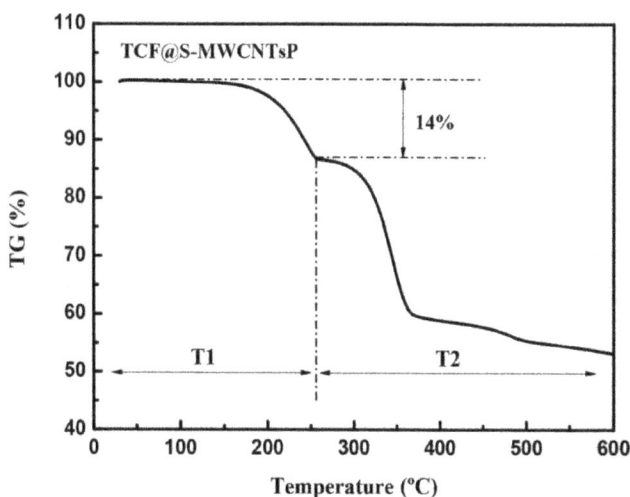

Figure 10. TG curve of the TCF@S-MWCNTsP electrode.

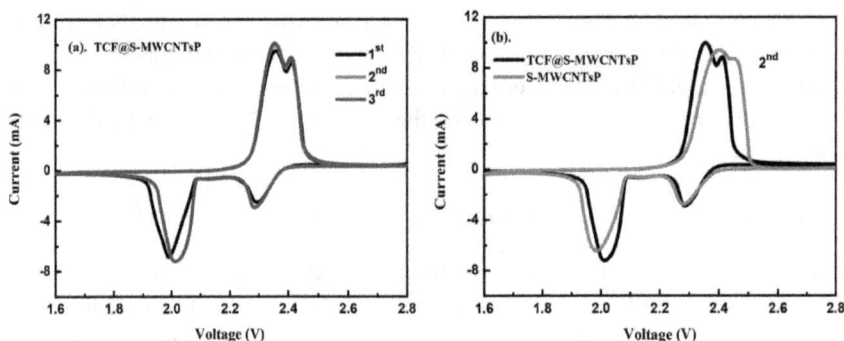

Figure 11. CV curves of Li-S batteries. (a) The first three cycles of CV profiles for TCF@S-MWCNTsP electrode. (b) The second cycle of CV profiles for TCF@S-MWCNTsP and S-MWCNTsP electrodes.

[44]. **Figure 11(b)** shows the second cycle of the CV plots for the two cathodes. There is a voltage shift between TCF@S-MWCNTs and S-MWCNTs electrodes other than the shape difference. This proves that using MWCNTsP as a current collector is beneficial to improve the electrochemical performance of a Li-S batteries. In addition, the TCF@S-MWCNTs electrode shows a higher voltage value at the reduction peaks and the oxidation peaks have a lower voltage value. This shows that the TCF@S-MWCNTs electrode has a higher discharge platform, which is conducive to improving the specific capacity and suppressing the shuttle effect. Meanwhile, the sharp reduction and oxidation peaks are also clear evidence of high reactivity of sulfur in the TCF@S-MWCNTs electrode. These results suggested that the electrode with MWCNTs transparent conducting film can inhibit effectively the shuttle effect and anchored PSs.

4.2.4. Constant current charge and discharge

Figure 12(a) shows the galvanostatic discharge–charge voltage profiles of S-MWCNTsP electrode at current rate 0.2 C (1 C = 1675 mAh/g) in the potential range from 1.6 to 2.8 V. It can be seen that there are two discharge plateaus for different rate at 2.3 and 2.1 V, respectively. But, only 2.1 V plateau exhibited a longer flat range which was ascribed to conformation of short-chain sulfides of Li_2S_2 and Li_2S. After 20 cycles, the discharge capacity also faded to 968 mAh/g from initial 1282 mAh/g. Another distinct characteristic is that Coulombic efficiency faded to 90.0% from initial 98.5%. It was considered the low Coulombic efficiency after 20 cycles was resulted from dissolution of long-chain PSs in electrolyte and subsequent migration and deposition on lithium anode. **Figure 12(b)** shows the galvanostatic discharge-charge voltage profiles of TCF@S-MWCNTsP electrode at current rate 0.2 C. The galvanostatic charge-discharge curves displayed a similar profile as S-MWCNTsP electrode. But, both the voltage plateaus of 2.3 and 2.1 V all exhibited longer flat range than ones of S-MWCNTsP electrode, indicating an excellent potential stability. The Coulombic efficiency and discharge capacity after 20 cycles reached 94.8% and 1028 mAh/g, respectively. This indicates that electrode with MWCNTs transparent conducting film can increase discharge capacity and Coulombic efficiency, inhibit shuttle effect.

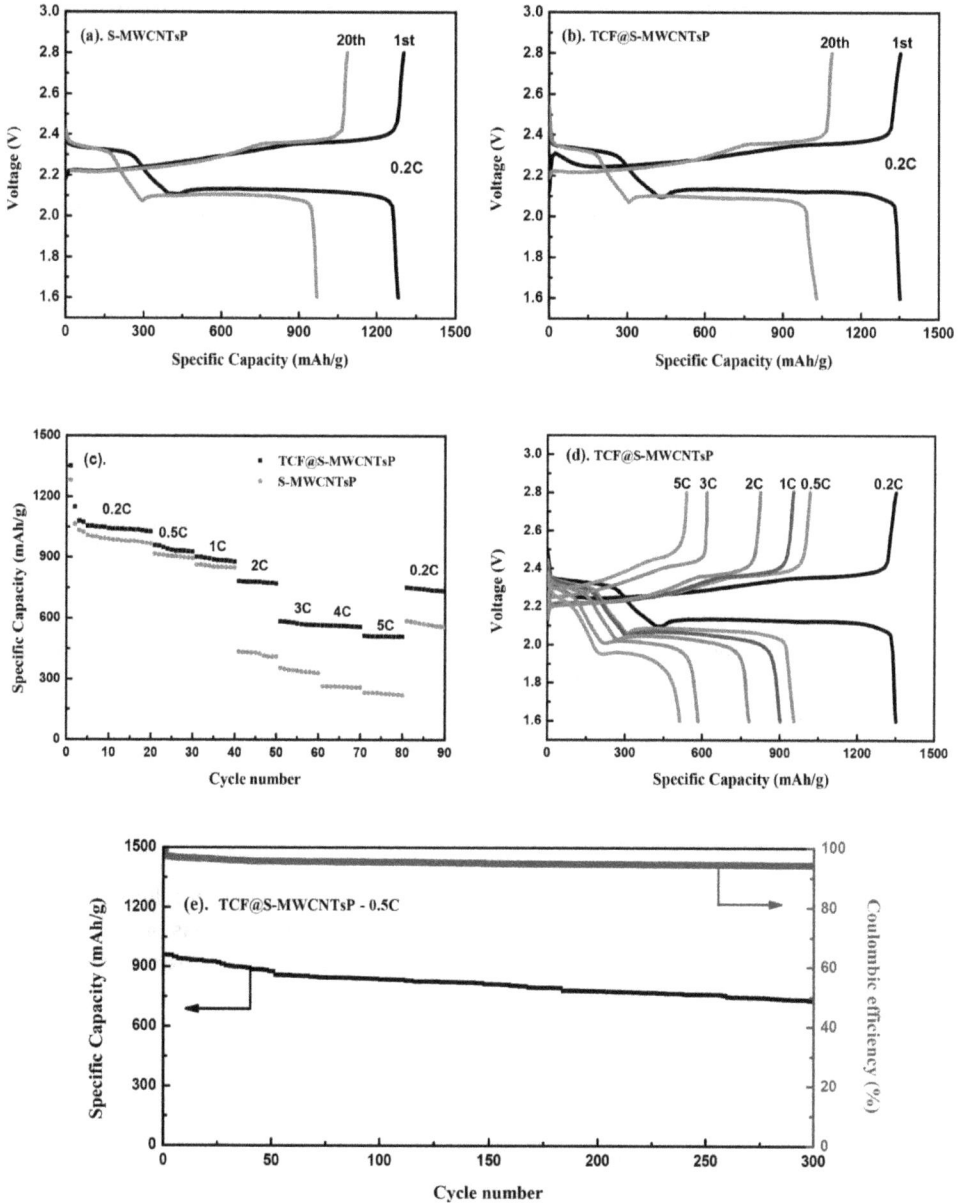

Figure 12. Electrochemical performance of Li-S batteries. Galvanostatic charge-discharge profiles of the (a) S-MWCNTsP and (b) TCF@S-MWCNTsP electrodes at 0.2 C. (c) The rate performance of TCF@S-MWCNTsP and S-MWCNTsP electrodes. (d) Galvanostatic charge-discharge profiles of the TCF@S-MWCNTsP electrode at various rates. (e) Long cycling performance of TCF@S-MWCNTsP electrode at 0.5 C.

Then the electrochemical performance of a Li-S cell was tested by galvanostatic discharge and charge from 1.6 to 2.8 V (versus Li/Li⁺) at different current rates, and the areal mass loading of sulfur in the cathode is controlled to be approximately 2.4 mg/cm². **Figure 12(c)** exhibits the rate performance of the two cathodes ranging from 0.2 to 5 C. Compared with the S-MWCNTsP and TCF@S-MWCNTsP cells, the TCF@S-MWCNTsP cell delivers a much higher initial capacity of 1352 mAh/g at the rate of 0.2 C, followed by a subsequent slow decrease to 960, 902, and 782 mAh/g at rates of 0.5, 1, and 2 C, respectively. In addition, at higher rates of 3 and 5 C, a reversible capacity of 584 and 513 mAh/g can still be achieved. When suddenly switching back to the initial starting rate of 0.2 C, the original capacity was recovered, indicating the excellent reversible capacity of the TCF@S-MWCNTsP cell at various rates. These results indicate that the electrode with MWCNTs transparent conducting film is conducive to immobilizing sulfur and alleviating the dissolution of polysulfides. The charge-discharge curves of the TCF@S-MWCNTsP electrode at various current rates (0.2–5 C) are illustrated in **Figure 12(d)**. All the discharge curves exhibit two typical plateaus, which are well consistent with the CV results. The charge-discharge voltage plateaus remain stable during the prolonged cycles, indicating an excellent potential stability. Additionally, the charge-discharge curves of the TCF@S-MWCNTsP electrode show also high Coulombic efficiency.

Long-term cycling stability with high-capacity retention is crucial for the practical application of Li – S batteries. **Figure 12(e)** shows the cycling performance of the TCF@S-MWCNTsP electrode at 0.5 C for 300 cycles. The electrode with MWCNTs transparent conducting film delivers a high initial reversible capacity of 960 mAh/g, and the capacity remains at 730 mAh/g after 300 cycles with stabilized coulombic efficiency above 94.2%, corresponding to a capacity retention of 76% and slow capacity decay rate of 0.08% per cycle. Additionally, the TCF@S-MWCNTsP electrode had the high Coulombic efficiency over 300 cycles, proving that the electrode with MWCNTs transparent conducting film can effectively suppress the notorious shuttle effect and improve the cycling stability.

4.2.5. Electrochemical impedance spectroscopy

The role of the electrode with MWCNTs transparent conducting film in Li-S batteries was further probed by electrochemical impedance spectroscopy (EIS). Nyquist plots of the two cells impedance before cycles are shown in **Figure 13(a)**. In the high-frequency region, the intercept of the impedance curve on the x-axis corresponds to the electrolyte resistance (Re). In the middle-frequency region, the semicircle arises from the charge transfer resistance (Rct), which represents the charge-transfer process at the interface between the electrolyte and electrode. In the low-frequency region, an inclined line denotes the Warburg resistance (Wo), which is related with mass transfer processes [45, 46]. The TCF@S-MWCNTsP electrode has the lower Rct value, indicating a low resistance caused by the entrapment of the dissolved PSs and both good electrolyte infiltration and charge transport. After 90 cycles, the Rct of S-MWCNTsP and TCF@S-MWCNTsP electrodes all demonstrates a decline as shown in **Figure 13(b)**. This can be ascribed to good electrolyte infiltration and the dramatic improvement of electronic and ionic conductivity due to the unique porous conductive interlinked structure of MWCNTsP. But the Re of both electrodes all demonstrates a rise. The increase of Re is caused by the dissolution and diffusion of PSs into the electrolyte. The lower Re value of the TCF@S-MWCNTsP electrode can be attributed to the use of the MWCNTs transparent conducting film framework can anchor PSs effectively and suppressing shuttle effects. Simultaneously, the electrode with

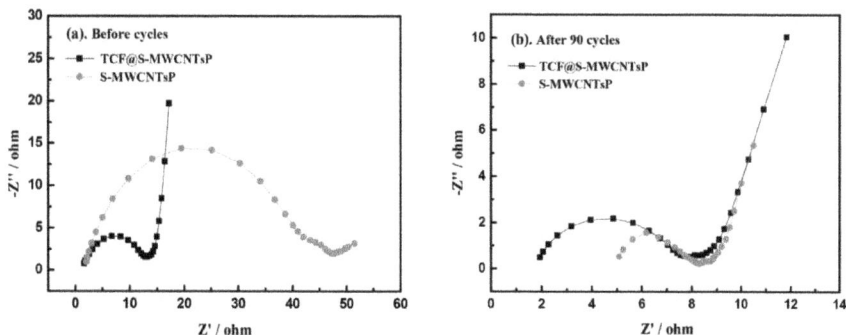

Figure 13. Electrochemical impedance spectroscopy of the two electrodes (a) before and (b) after cycles.

MWCNTs transparent conducting film could effectively reuse the dissolved active materials and mitigate surface aggregation, thus providing better performance.

5. Conclusions

The surface of R-MWCNTs has irregular arrangement of carbon atoms and much defects, which are more favorable for the adsorption of dispersant molecules. The dispersion effect and stability of R-MWCNTs are better than those of G-MWCNTs conductive fluids. With the increase of the dispersant TNADIS mass fraction, the dispersion effect of MWCNTs is getting better and better. When added to 0.05 wt.% dispersant, the dispersion effect of MWCNTs is best, and then the dispersive effect becomes worse as the dispersant TNADIS increases. With the increase of the quality of the dispersant TNADIS, the density of the MWCNTs precipitates becomes larger and larger, and the excess dispersant makes the MWCNTs agglomerate more and more closely. The conductivity of the conductive liquid with 0.05% dispersant was very high. After standing for 5 months, there was no obvious precipitation and the Tyndall effect of the colloid was still significantly maintained. Transparent conductive films prepared from conductive liquids of G-MWCNTs have better conductivity. After the three layers of spin-coating, the electrical conductivity of G-MWCNTs film surpassed three times than R-MWCNTs.

In summary, the designed electrode with MWCNTs transparent conducting film (TCF@S-MWCNTsP) showed significantly enhanced improvements in capacity retention and long-term cycle stability in Li – S batteries. The synergistic effect of them contributed to good rate performance, high capacity and excellent Coulombic efficiency. The areal sulfur mass loading of electrode is controlled to be above 2.4 mg/cm^2, a high discharge specific capacity of 1352 mAh/g can be delivered at 0.2 C with a Coulombic efficiency of 100%. Notably, the TCF@S-MWCNTsP electrode displayed a long cycling performance with only 0.08% capacity decay per cycle over 300 cycles at 0.5 C. The improved performance is ascribed to the trapping capability of the 3D configuration electrode to reutilize the dissolved polysulfides and the reduction of charge transfer impedance of the electrode. We believe that this attempt gives new insights on the cathode design for achieving high performance Li-S batteries.

Author details

Xiaogang Sun*, Jie Wang, Wei Chen, Xu Li, Manyuan Cai, Long Chen, Zhiwen Qiu, Yapan Huang, Chengcheng Wei, Hao Hu and Guodong Liang

*Address all correspondence to: xiaogangsun@163.com

Nanchang University, Nanchang, China

References

[1] Sheng L, Jin A, Yu L, et al. A simple and universal method for fabricating linear carbon chains in multiwalled carbon nanotubes. Materials Letters. 2012;**81**(3):222-224

[2] Ghosh K, Kumar M, Maruyama T, et al. Tailoring the field emission property of nitrogen-doped carbon nanotubes by controlling the graphitic/pyridinic substitution. Carbon. 2010;**48**(1):191-200

[3] Wang H, Li Z, Ghosh K, et al. Synthesis of double-walled carbon nanotube films and their field emission properties. Carbon. 2010;**48**(10):2882-2889

[4] Shinke K, Ando K, Koyama T, et al. Properties of various carbon nanomaterial surfaces in bilirubin adsorption. Colloids and Surfaces. B, Biointerfaces. 2010;**77**(1):18-21

[5] Ghosh K, Kumar M, Maruyama T, et al. Microstructural, electron-spectroscopic and field-emission studies of carbon nitride nanotubes grown from cage-like and linear carbon sources. Carbon. 2009;**47**:1565-1575

[6] Andoa K, Shinkea K, Yamadaa S, Koyamab T, et al. Fabrication of carbon nanotube sheets and their bilirubin adsorption capacity. Colloids and Surfaces. B, Biointerfaces. 2009;**71**(2):255-259

[7] Jiang X, Lu C, Song J, et al. Preparation of multi-walled carbon nanotubes acetone solution and study of dispersion stability. Insulating Materials. 2011;**44**(4):9-12

[8] Qi X. Dispersion of carbon nanotubes in aqueous solution with cationic surfactant CTAB. Journal of Inorganic Materials. 2007;**22**(6):1122-1126

[9] Kumar M, Ando Y. Camphor-a botanical precursor producing garden of carbon nanotubes. Diamond & Related Materials. 2003;**12**(5):998-1002

[10] Shimizu T, Abe H, Ando A, et al. Electric transport measurement of a multi-walled carbon nanotube in scanning transmission electron microscope. Physica E: Low-dimensional Systems and Nanostructures. 2004;**24**:37-41

[11] Taniguchi M, Nagao H, Hiramatsu M, et al. Preparation of dense carbon nanotube film using microwave plasma-enhanced chemical vapor deposition. Diamond & Related Materials. 2005;**14**(1):855-858

[12] Endo M, Takeuchi K, Hiraoka T, et al. Stacking nature of graphene layers in carbon nanotubes and nanofibres. Journal of Physics & Chemistry of Solids. 1997;**58**(11):1707-1712

[13] Endo M, Kim C, Karaki T, et al. Structural characterization of milled mesophase pitch-based carbon fibers. Carbon. 1998;**36**(11):1633-1641

[14] Endo M, Kim YA, Hayashi T, et al. Microstructural changes induced in "stacked cup" carbon nanofibers by heat treatment. Carbon. 2003;**41**(7):1941-1947

[15] Wakiwaka H, Kodani M, Endo M, et al. Non-contact measurement of CNT compounding ratio in composite material by eddy current method. Sensors & Actuators A Physical. 2006;**129**(1):235-238

[16] Xie JL, Chen XY, Jin H, et al. Synthesis and characterization of stable graphene colloidal dispersions. Journal of Functional Materials. 2014;**45**(12):12108-12112

[17] Ma P, Siddiqui NA, Marom G, et al. Dispersion and functionalization of carbon nanotubes for polymer-based nanocomposites: A review. Composites Part A Applied Science & Manufacturing. 2010;**41**(10):1345-1367

[18] Gauthier M, Reyter D, Mazouzi D, et al. From Si wafers to cheap and efficient Si electrodes for Li-ion batteries. Journal of Power Sources. 2014;**256**(12):32-36

[19] Yang Y, Zheng G, Cui Y. Nanostructured sulfur cathodes. Chemical Society Reviews. 2013;**42**(7):3018-3032

[20] Liang J, Sun ZH, Li F, et al. Carbon materials for Li–S batteries: Functional evolution and performance improvement. Energy Storage Materials. 2016;**2**:76-106

[21] Manthiram A, Fu Y, Chung SH, et al. Rechargeable lithium-sulfur batteries. Chemical Reviews. 2014;**114**(23):11751

[22] Evers S, Nazar LF. New approaches for high energy density lithium–sulfur battery cathodes. Accounts of Chemical Research. 2013;**46**(5):1135

[23] Sen X et al. Smaller sulfur molecules promise better lithium–sulfur batteries. Journal of the American Chemical Society. 2012;**134**(45):18510-18513

[24] Yu M, Ma J, Song H, et al. Atomic layer deposited TiO_2 on a nitrogen-doped graphene/sulfur electrode for high performance lithium–sulfur batteries. Energy & Environmental Science. 2016;**9**(4):1495-1503

[25] Zhang Q, Wang Y, Zhi WS, et al. Understanding the anchoring effect of two-dimensional layered materials for lithium–sulfur batteries. Nano Letters. 2015;**15**(6):3780

[26] Manthiram A, Chung SH, Zu C. Lithium-sulfur batteries: Progress and prospects. Advanced Materials. 2015;**27**(12):1980-2006

[27] Pang Q, Kundu D, Cuisinier M, et al. Surface-enhanced redox chemistry of polysulphides on a metallic and polar host for lithium-sulphur batteries. Nature Communications. 2014;**5**:4759

[28] Mikhaylik YV, Akridge JR. Polysulfide shuttle study in the Li/S battery system. Journal of the Electrochemical Society. 2004;**151**(151):A1969-A1976

[29] Busche MR, Adelhelm P, Sommer H, et al. Systematical electrochemical study on the parasitic shuttle-effect inlithium-sulfur-cells at different temperatures and different rates. Journal of Power Sources. 2014;**259**:289-299

[30] Zhang J, Hu H, Li Z, et al. Double-shelled nanocages with cobalt hydroxide inner shell and layered double hydroxides outer shell as high-efficiency polysulfide mediator for lithium-sulfur batteries. Angewandte Chemie. 2016;**55**(12):3982

[31] Hofmann AF, Fronczek DN, Bessler WG. Mechanistic modeling of polysulfide shuttle and capacity loss in lithium–sulfur batteries. Journal of Power Sources. 2014;**259**(259):300-310

[32] Zhou G, Pei S, Li L, et al. A graphene–pure-sulfur sandwich structure for ultrafast, long-life lithium–sulfur batteries. Advanced Materials. 2014;**26**(4):625-631

[33] Huang JQ, Zhang Q, Wei F. Multi-functional separator/interlayer system for high-stable lithium-sulfur batteries: Progress and prospects. Energy Storage Materials. 2015;**1**:127-145

[34] Ko W, Su J, Guo C, et al. Highly conductive, transparent flexible films based on open rings of multi-walled carbon nanotubes. Thin Solid Films. 2011;**519**(22):7717-7722

[35] Shin E, Jeong G. Fabrication of transparent, flexible and conductive films using as-grown few-walled carbon nanotubes. Current Applied Physics. 2011;**11**(4):S73-S77

[36] Xiao G, Tao Y, Lu J, et al. Highly conductive and transparent carbon nanotube composite thin films deposited on polyethylene terephthalate solution dipping. Thin Solid Films. 2010;**518**(10):2822-2824

[37] Jung R, Kim H, Kim Y, et al. Electrically conductive transparent papers using multiwalled carbon nanotubes. Journal of Polymer Science Part B Polymer Physics. 2008;**46**(12):1235-1242

[38] Jo S, Lee Y, Yang J, et al. Carbon nanotube-based flexible transparent electrode films hybridized with self-assembling PEDOT. Synthetic Metals. 2012;**162**(13-14):1279-1284

[39] Zhang J, Wen X, Song Q, et al. Research advances in fabrication technologies of transparent conducting carbon nanotube films. Materials Review. 2011;**25**(5):124-129

[40] Ning J, Zhi LJ. Advances in flexible transparent conductive films based on carbon nanomaterials. Chinese Science Bulletin. 2014;**59**:3313-3321

[41] SH L, CC T, CC M, et al. Highly transparent and conductive thin films fabricated with nano-silver/double-walled carbon nanotube composites. Journal of Colloid and Interface Science. 2011;**364**(1):1-9

[42] Zhao MQ, Zhang Q, Huang JQ, et al. Unstacked double-layer templated graphene for high-rate lithium-sulphur batteries. Nature Communications. 2014;**5**(5):3410

[43] Salem HA, Babu G, Rao CV, et al. Electrocatalytic polysulfide traps for controlling redox shuttle process of Li–S batteries. Journal of the American Chemical Society. 2015;**137**(36):11542

[44] Yijuan L et al. A honeycomb-like Co@N–C composite for ultrahigh sulfur loading Li–S batteries. ACS Nano. 2017;**11**(11):11417-11424

[45] Tao H, Mukoyama D, Nara H, et al. Electrochemical impedance spectroscopy analysis for lithium-ion battery using $Li_4Ti_5O_{12}$ anode. Journal of Power Sources. 2013;**222**(2):442-447

[46] Li S, Ren G, Hoque MNF, et al. Carbonized cellulose paper as an effective interlayer in lithium-sulfur batteries. Applied Surface Science. 2017;**396**:637-643

Highly Conductive and Carbon Nanotube Activated Transparent Thin-film

Chapter 4

Carbon Nanotube-Activated Thin Film Transparent Conductor Applications

Iskandar Yahya, Seri Mastura Mustaza and
Huda Abdullah

Additional information is available at the end of the chapter

http://dx.doi.org/10.5772/intechopen.79367

Abstract

Carbon nanotubes are an exciting nanomaterial system that exhibit exceptional mechanical and electrical properties. Due to its small diameter of ∼1 nm and high aspect ratio in order of 10^3, they can readily form interconnected conducting network of continuous film with high transparency. Transparent conductor based on carbon nanotube can be grown directly into a thin-film structure, or can be processed after the growth process. Post-growth arrangement of carbon nanotube into transparent conducting thin films can be achieved by several methods. Most of the methods involve solution-processed approach, while dry-processed approach is also possible. This chapter presents a comprehensive review and methods for fabricating transparent carbon nanotube-activated thin film, which generally demonstrate high conductivity and mechanical flexibility. Comparison on the optical and electrical performance of the carbon nanotube-activated transparent conductors fabricated via different methods is presented in the chapter.

Keywords: carbon nanotube, transparent, conductor, thin films, fabrication, methods

1. Introduction

The advent of mobile and interactive devices has seen tremendous increase in demand for transparent conductors. The commonly used material as transparent conductor such as indium tin oxide (ITO) is scares, expensive, brittle and suffers from thermal-related degradation. There is a need for alternative material system which is cheaper and readily available to complement or even replace ITO in the future. Such a material system is carbon nanotube-activated thin film transparent conductor.

Carbon nanotube-activated thin film transparent conductor exhibits excellent electrical conductance, high carrier mobility, high flexibility, and environmental robustness. Furthermore, coupled with a variety of simple fabrication methods, this material is suitable for emerging technology applications such as photovoltaic electrodes, organic light emitting diodes (OLEDs), touch panels, field emitters, smart windows, and smart fabrics. The easiest route for fabricating carbon nanotube-activated thin film transparent conductor is via simple spin-coating method.

Recent work utilizing the multi-walled carbon nanotubes has shown to reduce sheet resistance via nanotube length optimization [1]. Double-walled carbon nanotubes can also be used to produce the thin film transparent conductors via dip coating method and has been shown to be superior to that of multi-walled carbon nanotubes with low sheet resistance of $\sim$134 $\Omega/\square$ or lower and 99% transmittance at 550 nm [2]. Single-walled carbon nanotubes activated transparent conducting films on the other hand can be further doped with $HAuCl_4$ to enhance the optoelectronic performance of $\sim$40 $\Omega/\square$ sheet resistance and $\sim$90% transmittance [3].

A successful demonstration of carbon nanotube-activated thin film transparent conductor as a strain sensor application showed high sensing range of up to 400% and fast response of less than $\sim$98 ms with a transmittance of 80% at 550 nm [4]. The performance of carbon nanotube-activated thin film is also comparable to graphene-activated thin film. A recent demonstration of inverted perovskite solar cell application of carbon nanotube- versus graphene-activated thin film showed a power conversion efficiency of 12.8 and 14.2%, respectively [5]. It is also possible to integrate both carbon nanotube and graphene in a hybrid structure as the thin film to further enhance the optoelectronic and mechanical performances [6].

In this chapter, the basic theoretical background of carbon nanotubes is briefly explored, followed by their application as transparent conductors. In the following section, an in-depth discussion of selected transparent film fabrication methods such as dip coating, vacuum filtration, and Langmuir-Blodgett is carried out. The optical and electrical performances of films fabricated by the different fabrication methods are also compared.

2. Carbon nanotubes

2.1. Carbon nanotube properties

Carbon nanotube (CNT) is a type of carbon allotrope, also referred to as nanoallotropes due to their nanometer dimensions. It belongs to a graphemic nanostructure group consisting of densely packed hexagonal honeycomb sp^2 carbon crystal lattice, similar to graphene. CNTs structurally are hollowed tubular cylinder with wall made up of a hexagonal carbon honeycomb with a very high aspect ratio in order of thousands.

A single-walled carbon nanotube (SWCNT) is a type of CNT with a wall consisting of a single atomic layer of carbon atoms bonded together in a hexagonal honeycomb structure. It can be imagined as a monolayer of graphene rolled into a capped cylinder. However, the synthesis method is not as straightforward. Depending on the orientation of the rolling axis of the graphene layer, SWCNTs exhibit varying electronic properties, that is, varying bandgap and also varying diameter in the range of 0.7–3.0 nm. This varying tube properties, or species, based on the varying rolling axis orientation is denoted by the chiral number (n, m). Therefore, SWCNTs can be tuned to exhibit electronic properties that of metallic elements (conducting) or semiconducting with varying bandgap based on its chirality. Studies have shown that an (n, m) tube is metallic when $n = m$, or when $n - m = 3i$, where i is an integer, while semiconducting CNTs have $n - m \neq 3i$ [7].

Figure 1 shows examples of SWCNT structures. A zigzag SWCNT is a species when $m = 0$, an armchair SWCNT is a species, when $n = m$, and a chiral SWCNT is a species with any other combination of (n, m). A multi-walled carbon nanotube (MWCNT) is another type of CNT with a stack of multiple graphene layers forming the wall structure in a concentric fashion similar to a Russian doll. MWCNTs can have a diameter anywhere in the range of 3–20 nm. Theoretically, tube diameter in the excess of 100 nm is possible for MWCNT. CNTs exhibit extraordinary mechanical and electrical properties for a 1D quantum wire. In particular, CNTs have high electrical conductivity ($\rho \approx 10^{-6}$ Ω m) [8]. This high conductivity arises from the confinement of electrons, which allows one dimensional (1-D) electron movement in either direction along a single line and the requirements for energy and momentum conservation, resulting in reduced scattering processes [9]. Theoretically, in the absence of scattering, transport is ballistic with unit conductance $G = 2e^2/h$ with the resistance of 12.9 kΩ [10]. CNTs also have high thermal conductivity ($k = 1750$–$58,000$ W/mK) [11], high tensile strength (60 GPa) [12], and

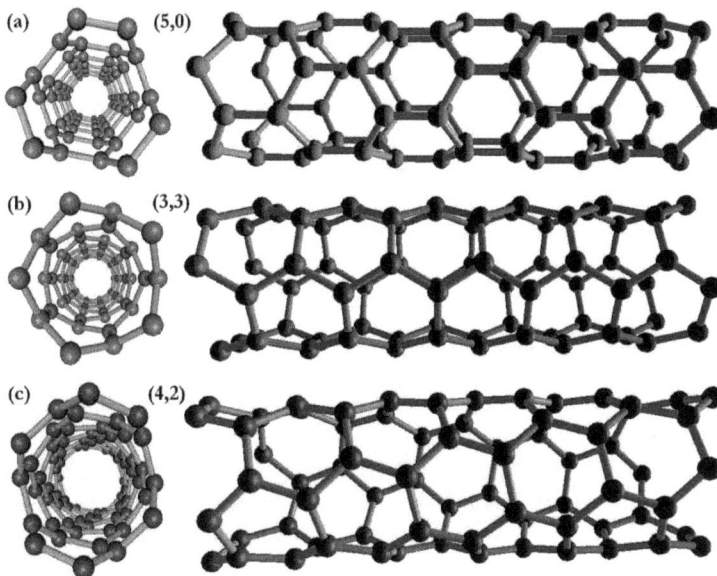

Figure 1. Graphical representation of different SWCNT physical structures from the front (left) and side perspectives (right). (a) A zigzag SWCNT with chiral number (5,0). (b) An armchair SWCNT with chiral number (3,3). (c) A chiral SWCNT with chiral number (4,2).

high Young's modulus (1 TPa) [13]. The closed structure at the edges and no dangling bonds in CNTs makes them chemically stable and inert. The C atoms in CNTs are completely covalently bonded, and therefore do not suffer from electromigration or atomic diffusion like metals, making them a very good electrical conductor that can sustain high current densities (>10^9 A/cm^2) [14, 15]. All these properties make CNTs one of few materials that have very good prospects of driving technology progress in future electronics and other applications.

Most of the applications of CNTs are due to its mechanical resilience and ballistic electric conductivity. The vast applications of CNTs are evident with different new findings being reported continually over the last two decades. In electronics, the most important applications are carbon nanotube field effect transistors (CNTFET), circuit interconnects, and transparent conducting films. New studies of CNTFETs led to many findings of its new uses; CNT power transistors, biosensors, electromagnetic wave sensors, gas sensors, memory elements, and transparent flat devices are some of the possible applications derived from a basic CNTFET operation.

Currently, the popular methods of mass producing CNTs are arc discharge, laser ablation, and chemical vapor deposition (CVD). Ebbesen and Ajayan in 1992 managed to demonstrate the growth and purification of MWCNT with high quality and quantity (gram level) by arc discharge [16]. Arc discharge technique involves generating an electric arc between two graphite electrodes with one of them filled with catalyst metal powder such as Fe, Ni, and Co, in a He gas atmosphere. In the laser ablation method, a laser is used to evaporate a graphite target that is mixed with catalyst powder [8]. Both arc discharge and laser ablation methods produce only up to 70% CNTs, while the rest are amorphous carbon and catalyst particles, making it necessary to carry out a post-growth purification step. The CVD process uses catalyst metal nanoparticles reacted in hydrocarbon gas at temperatures of 450–1100°C [17]. The catalyst nanoparticles will decompose and dissolve the hydrocarbon gas before precipitating out from its circumference as CNT [18]. Recently, the CVD method has gained much popularity because of the ability to control the morphology and the quality of the CNT that sparked different variants of the CVD methods such as plasma enhanced CVD (PE CVD) [19], high-pressure catalytic decomposition of carbon monoxide (HiPCO) [20], Co-MoCat process [21], and alcohol CVD [22].

2.2. Carbon nanotubes as transparent conductors

A lot of study has been done in applying CNTs as materials for transparent conductors [23]. The key advantages of CNT as transparent conductor relate to their outstanding structure and electronic properties. Structurally, CNTs have extremely small diameters of ~1 nm and high aspect ratio in the order of 10^3, making them readily to form intercalating assembly of continuous tube network as thin as a few nanometers. Being mechanically strong and flexible, a thickness of a few nanometers is enough to mechanically manipulate them into device structures. Electronically, CNTs show outstanding conductivity and current carrying capacity up to 1000 times better than Cu. Combining these two characteristics of CNT films, they are suitable and exciting material system for application in technologically relevant electronic devices requiring high conductivity and transparency toward visible light electromagnetic radiation and beyond.

Furthermore, SWCNTs can be metallic or semiconducting based on its chirality. Normally, synthesized SWCNTs will contain statistically a mixture of both types of tube with the ratio of metallic to semiconducting tube to be 1–3, approximately [9]. Ideally, transparent conducting films based on SWCNTs require tube composition of totally metallic tubes to maximize the total conductivity. Efforts to grow selectively metallic or semiconducting SWCNTs have remained as the holy grail of CNT electronics and have been carried out rigorously over last decade. It is possible to selectively enrich the growth of metallic [24] or semiconducting [25] SWCNTs. However, obtaining 100% selectivity is still not possible or at least still cannot be completely verified due to the limitations of the characterization methods available. In addition, selective growth of SWCNTs also requires specific growth platform or substrate, such as specifically oriented quartz substrates, which can pose additional problem in transferring and processing the tubes into thin transparent films.

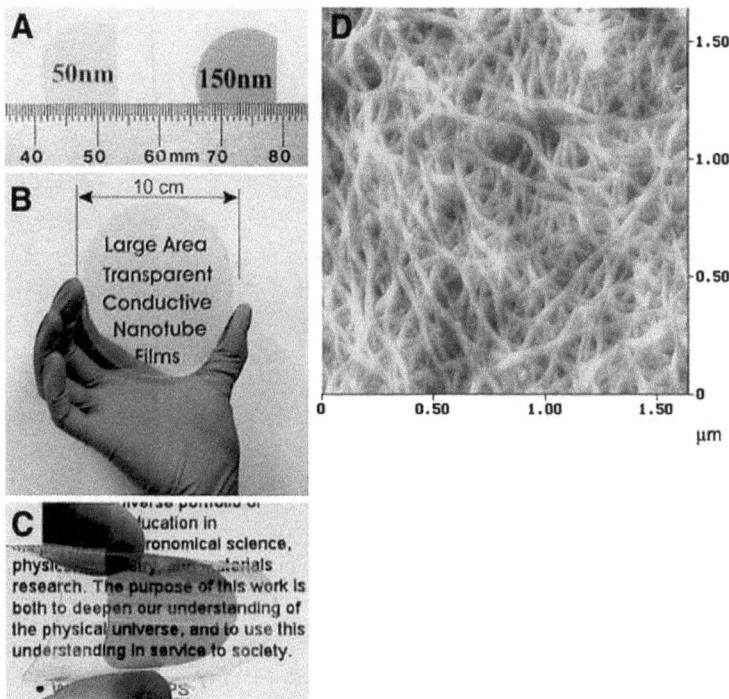

Figure 2. Transparent SWNCT films from A. G. Rinzler's group, taken from reference [33]. (a) Films of the indicated thickness on quartz substrates. (b) A large, 80-nm-thick film on a sapphire substrate 10 cm in diameter. (c) Flexed film on a Mylar sheet. (d) AFM image of a 150-nm-thick t-SWNT film surface (color scale: black to bright yellow, 30 nm). The text in (a)–(c) lies behind the films. From Ref. [33]. Reprinted with permission from AAAS.

Alternatively, as grown SWCNTs (mixed between metallic and semiconducting) can be used and subjected to post-growth separation based on electronic type, that is, metallic or semiconducting. Post-growth separation of SWCNT demonstrated includes dielectrophoresis [26, 27], gel electrophoresis [28], gel agarose chromatography [29, 30], density gradient ultracentrifugation [31], and among others. By enriching the SWCNT population with metallic tubes, the overall conductivity of the fabricated conducting thin films can be increased.

MWCNTs can also be applied to form thin transparent conducting films. Since MWCNTs exhibit metallic behavior, they can be directly processed without further post-growth separation, except for the standard purification step to remove impurities. However, it should be noted that only the outmost shell of the MWCNTs is involved in charge carrier conduction [32].

Because the fabricated CNT films can have a thickness between only a few nanometers up to the order of 100 nanometers, the transmittance can be as high as 95% in the 2–5 μm spectral band [33], which can rival the industry standard indium tin oxide (ITO) films. **Figure 2** shows some example SWCNT transparent thin films produced by vacuum filtration method demonstrated by Wu *et al.* of Zingler's group. The atomic force microscopy analysis (AFM) in **Figure 2D** shows individual SWCNTs randomly intercalated in a network to form an electrically conducting film.

3. Fabrication methods of CNT transparent conducting films

The techniques for forming continuous intercalating network of CNTs to form transparent films can be divided into two main approaches, which is dry processed and solution processed. Solution-processed approach involves dispersing CNTs into solutions, either H_2O based or solvent based. Dispersed CNTs in solution can then be directly deposited on a surface to form a thin film followed by a drying process. The deposition or film forming process can be done by a number of methods including vacuum filtration, dip coating, spin coating, spray coating, direct ink printing, Langmuir-Blodgett technique, and roll-to-roll, among others. Dry-processed approach, as the name suggests, does not involve any use of solution as dispersant for CNTs. The CNTs are either grown directly into a continuous film form or handled directly into films after the CNT synthesis step. In this chapter, both the dry-processed approach and solution-processed approach are discussed briefly.

3.1. Dry-processed approach

The dry-processed approach in producing transparent conducting CNT films that is discussed here involved two methods: pulling of vertically grown MWCNTs and direct film growth on quartz substrate.

In the first method, vertical MWCNT forests grown via catalytic CVD are pulled by hand from a sidewall using an adhesive strip, which was demonstrated by Zhang et al. [34]. The pulled MWCNTs from the sidewall will form bundles, which in turn will trap the MWCNTs in the row adjacent to it and pulling them out too, forming a continuous sheet of MWCNTs.

Depending on the size of the supporting substrate of the MWCNT forest, the dimension of a single continuous sheet in a single pull is ~5 cm wide and 100 cm long. The fastest pulling rate at which the MWCNTs can self-support itself without breaking is ~700 cm/min. However, the pulling technique only work with certain vertically grown MWCNT forest and the allowable pulling rate depends on the forest structure. The resulting sheets of MWCNT appear to be oriented parallel to the pulling direction. From a ~245-μm-high forest, the resulting sheet thickness is ~18 μm with an areal density of ~2.7 $\mu g/cm^2$, and a volumetric density of ~0.0015 g/cm^3 [34]. This indicates that in fact, the MWCNT sheets produced are electronically conducting and highly aligned aerogel. The thickness of the sheets can be reduced to ~50 nm, effectively increasing the tube density to ~0.5 g/cm^3, by laying them flat on a substrate and immersing it in a liquid where the surface tension will compress the aerogel into thin transparent conducting films. The resulting sheet resistivity was shown to be ~700 $\Omega/\square$ and transmittance of more than 85% [34].

In the second method, arrays of CNTs are grown parallel to the substrate plane, forming a thin continuous conducting film. It was shown that aligned SWCNTs are grown on mono-crystalline ST-cut quartz substrate by using ethanol as the carbon feedstock and Cu as the catalyst [35]. By using this growth method, the SWCNTs will align themselves along the X direction of the ST-cut mono-crystalline quartz surface, similar to the <100> plane direction in Si wafer. Depending on the density of the catalyst particles, growth of up to 50 SWCNTs per μm can be achieved, with tube length of up to a few millimeters [35]. It is also possible to apply the same growth method to selectively grown semiconducting SWCNTs. This can be done by selectively etching random growing SWCNTs and mixing water vapor with the carbon feedstock to selectively grown SWCNTs of a particular electronic type, in this case semiconducting [25]. Selective growth of metallic SWCNTs should be possible using the same principle by optimizing the correct water vapor to carbon feedstock ratio, or by using a different carbon feedstock. However, it was also shown that aligned SWCNTs do not make high performance transparent conducting film, and that randomly oriented network of SWCNTs is preferred [36]. By using ferritin as the catalyst precursor instead, SWCNTs grown on quartz substrate were parallel to the surface plane but highly random in growth direction. Higher density of SWCNTs can be obtained with resulting sheet resistance of ~7.1 k$\Omega/\square$ and transmittance of up to ~98.4%. **Figure 3** shows the scanning electron microscope (SEM) images of randomly oriented (**Figure 3a**) thin film of SWCNTs and aligned array of SWCNTs (**Figure 3b**) grown on the quartz substrate. It is also possible to directly grow CNT-graphene hybrid materials to form transparent conducting films with a sheet resistance of ~450 $\Omega/\square$ and with transmittance of up to 86% at 550 nm [37].

3.2. Solution-processed approach

Solution-processed approach involved dispersing CNTs into either an aqueous-based solutions or solvents, followed by the deposition process, which can be done through several techniques. The purpose of this dispersion step is to separate CNTs from bundles into isolated suspension of nanomaterials in a solution for easy manipulation and deposition into thin films with controllable parameters. It is known that CNTs tend to bundle up or aggregate together

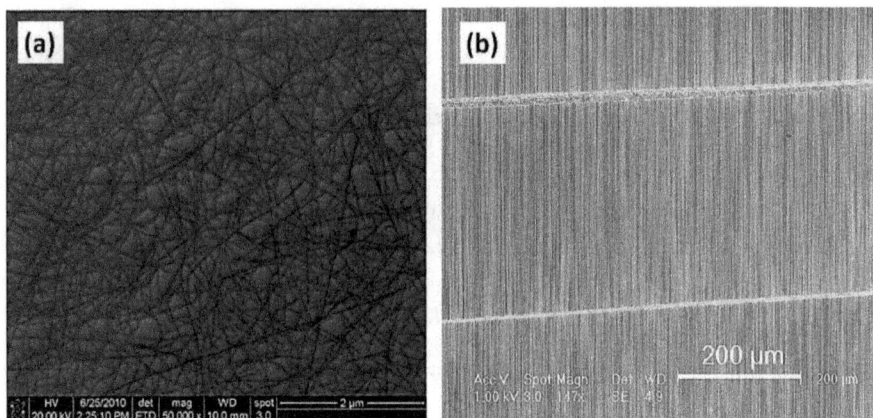

Figure 3. Scanning electron microscopy of SWCNTs directly grown into thin films on quartz substrate. (a) Randomly oriented SWCNTs adapted from Ref. [36], with permission from Elsevier. (b) Aligned array of SWCNTs adapted with permission from Ref. [35]. Copyright (2008) American Chemical Society.

due to the intrinsic electrostatics, especially when they are immersed in a solution. Therefore, to improve the dispersion or "solubility" of CNTs in solutions, particular choice of solution with added surfactants or organic solvents can be used.

There are three main solution types that can be used as dispersant solution: (a) organic solvents or superacids [38], (b) aqueous solution with added dispersing agent or surfactants, and (c) adding functional groups to the CNTs outer wall to counter the intermolecular attraction.

For organic solvents, commonly used solvents are dimethylformamide (DMF), N-methyl-2-pyrrolidone (NMP), toluene, chloroform, dichlorobenzene, and ethylene dichloride (EDC). However, the disadvantages of using organic solvents include low dispersion density of ~0.1 mg/mL, very high or very low boiling points of the solvents, and the tendency for the solvents to dissolve other materials involved in the deposition process such as plastics and polymers. Furthermore, depending on the organic solvent use, the dispersed CNTs tend to rebundle or aggregate again if left suspended for a certain period of time.

CNTs can also be dispersed in an aqueous solution mixed with surfactants that acts as surface active agents, which can assist in the dispersion of hydrophobic CNTs due to their amphiphilic properties [39]. Commonly used surfactants are sodium cholate (SC), sodium dodecyl sulfate (SDS), sodium deoxycholate (DOC), and sodium dodecylbenzenesulfonates (SDBS), among others. Nonionic detergent-based surfactants can also be used such as Triton X-100 (Sigma-Aldrich) and Tween 20 (Sigma-Aldrich). Other nonsurfactant solubilization agents that can be used include DNA, cellulose derivatives, porphyrins, starches, polysaccharides, and polymers. The effectiveness of the solubilizing agents and surfactants vary and depends on their head-group charge, inclusion of aromatic ring such as benzene, and also hydrophobic tail.

Another dispersion technique available is functionalization of the exposed CNT carbon atoms with other molecules via covalent bonding to promote affinity between the CNTs and the solution used. Functionalization also negates or reduces the effect of CNTs' electrostatic attraction between them [40]. Most functionalization materials used are acid based. Although the resulting dispersion of CNTs will have significantly higher tube concentrations compared to the other two approaches, the conductivity of the produced films may suffer from electrical degradation due to the process induced defects to the CNT sp2 structure.

All the dispersion techniques described here require assistance in the form of physical agitation. High power agitation tools such as ultrasonic bath, homogenizer, and physical tip ultrasonic probe may have to be used during the dispersion process. For example, dispersing CNTs in organic solvents requires constant agitation either via ultrasonic bath or ultrasonic tip probe to exfoliate individual CNT from the bundles. Similar technique is also required for CNT in aqueous-based solution with surfactants. Furthermore, this physical agitation also helps to break down impurity particles such as amorphous carbon and catalysts, which can be removed by ultracentrifugation. The heavier catalyst and amorphous carbon particles will be forced to precipitate, while the CNTs remain suspended in the solution as a direct effect of dispersion.

3.3. Film deposition methods

As described earlier, based on the solution-processed approach, there are a few film deposition techniques or methods that can be employed. Here, we will discuss the most commonly used methods, that is, dip coating, spraying, spin coating, vacuum filtration, ink-jet printing, and Langmuir-Blodgett techniques.

3.3.1. Dip coating technique

Dip coating method involves immersing a substrate vertically into a solution with dispersed CNTs and retracting it slowly [2]. Normally, the retracting speed is controlled (typically between 1 and 10 mm/min) to promote CNT adhesion to the substrate layer as the solution near the meniscus evaporates as denoted in **Figure 4**. The dip coating cycle is normally repeated several times to obtain reasonably homogenous and continuous transparent film of several nanometers thick. The thickness of the film can be increased by increasing the number of dip coating cycle to up to 400 times [41]. However, there is a limit to the thickness of the resulting CNT film as it reaches saturation, whereby any additional dip coating cycle will not

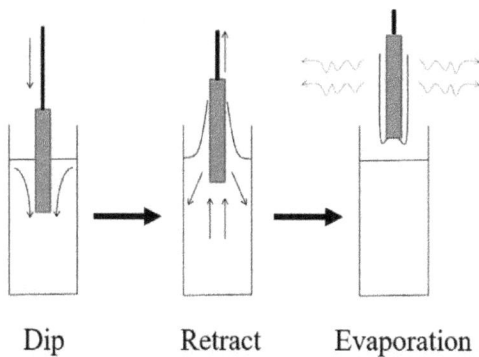

Dip   Retract   Evaporation

Figure 4. Dip coating technique. Adapted from Ref. [48] with permission from Sains Malaysiana, Universiti Kebangsaan Malaysia.

increase the thickness. For a typical CNT dispersed in organic solvent or aqueous solution with surfactant, maximum film thickness is between 50 and 100 nm, depending on the dispersion concentration and type of CNTs. The typical sheet resistance of a dip-coated CNT film was shown to be $\sim$300 $\Omega/\square$ and transmittance of $\sim$85% [41].

3.3.2. Spraying deposition technique

The spraying deposition technique involves blowing compressed air through the droplets of CNT dispersion. Typically used tool is the artist air-brush, which has been shown to produce thin, uniform film of CNTs on the substrate surface [41, 42]. The CNTs used are dispersed in aqueous solution with added surfactants such as SDS or SDBS. In this technique, water in the aqueous solution will dry up almost immediately as the spray droplets touch the substrate during spraying. This negate the effect of water droplet surface tension pulling CNTs to clump together during drying as in the dip-coating or spin-coating methods, dubbed as the coffee ring effect. It is also important to control the spraying parameters to make sure that the solution in the CNT dispersion dries on contact and not before. Premature drying will lead to dry CNTs that do not form good electrical contacts within the film network. However, it is difficult to control the drying of the solution and to estimate when complete drying has been achieved. It was shown that at film thickness of $\sim$26 nm, the resulting sheet resistance was measured to be $\sim$57 $\Omega/\square$ (conductivity, $\sigma \sim$ 6704 S/cm) and transmittance of $\sim$65%. Although, spray deposited CNT film shows high conductivity, the transmittance is relatively lower. It was also shown that comparing CNT films with sheet resistance of $\sim$475 $\Omega/\square$ deposited via dip coating and spray coating, the transmittance is $\sim$83 and $\sim$73%, respectively [41].

3.3.3. Spin-coating technique

The spin-coating technique employs similar tools for photolithography. A spin coater is used to spin a substrate after a deposition of CNT dispersion for even distribution [1]. This technique relies on the CNTs to form physisorption or adhesion to the substrate surface as the solution is pushed out due to the centripetal force. It may be essential to first coat the substrate with silane-based self-assembled monolayer (ASM) to promote adhesion between the CNT and the substrate surface. The spin cycle has to be repeated several times until a continuous homogenous CNT film is obtained. This method is simple but uses a lot of materials because most of the CNTs deposited during spinning will be washed out along with the solution. Consequently, this method is more suitable to be used with CNTs that are dispersed in organic solvents with low boiling points such as EDC and chloroform, where the solvents will dry up quickly during spinning and promote adhesion between the CNT and the substrate. The resulting CNT film will exhibit varying characteristics. One study has shown that by using CNT in EDC for spin coating, a film thickness of 24 nm was produced with a sheet resistance of $\sim$128 $\Omega/\square$ ($\sigma \sim$ 4629 S/cm) and transmittance at 550 nm of $\sim$90%.

3.3.4. Vacuum filtration technique

The vacuum filtration technique is straight forward and easy, as described in the detail in [33, 38]. Purified dispersion of CNTs in aqueous solution with surfactant or in superacids is first obtained as previously described. Here, any surfactant can be used. Basically, a diluted CNT dispersed in solution is filtered through a membrane filter using a vacuum filtration apparatus. The filtered CNT on the membrane will form a continuous thin film of CNT. It is recommended to use mixed cellulose ester (MCE) membrane filter with ~0.2 μm pore size [33]. Effectively, the size of the CNT film produced is limited by the vacuum filtration apparatus size or the filtering membrane size.

The following are essential steps to follow in order to obtain good transparent CNT film. The vacuum filtration apparatus is first equipped with the suitable membrane filter and is turned on. Solution of dispersed CNT is slowly and carefully pipetted into the filter reservoir funnel, while preventing the formation of bubbles. Any agitation due to fast water flow movements or bubbles will affect the continuity and homogeneity of the resulting CNT film. Additional CNT in solution can be added to achieve desired film thickness. After all the solutions have been filtered, the vacuum pump is left on for ~20 minutes to aid drying of the membrane filter. Deionized water is then pipetted inside the filter reservoir to wash out any residual surfactants until none is visible. After further drying, the CNT film on the membrane filter is then ready to be transferred onto a substrate.

To transfer the CNT film onto another substrate, it must be first removed from the filter membrane. The CNT film cannot be peeled off of the membrane filter due to the strong adhesion between the CNT and the membrane. Therefore, the membrane must be removed by dissolving it in organic solvents such as acetone. The membrane can be carefully immersed in an acetone bath until all visible traces of MCE membrane is gone. The floating CNT film can then be picked up using the substrate or using a sieve before transferring it onto a substrate. It is also possible to use poly(ethylene terephthalate) (PET) as the substrate. Subsequent washing of the film on substrate with acetone or other suitable solvent is repeated to remove the remaining traces of membrane filter. Annealing of the substrate at ~90°C can help improve adhesion between the CNT film and the substrate. It is possible to use flexible transparent material as the substrate, as depicted in **Figure 2**, which is the example of films produced via vacuum filtration method. **Figure 5a** shows CNT filtered on a membrane, and **Figure 5b** shows the CNT transparent film on glass substrate after membrane removal and film transfer.

Transparent CNT films produced by vacuum filtration method show promising conductivity to transmittance performance. Results showed an average conductivity, σ, of ~2000 S/cm with corresponding transmittance of ~93% at 550 nm wavelength [38]. Others have reported films of 50-nm thick to exhibit >90% transmittance and sheet resistance of 30 Ω/□ [33].

Vacuum filtration method is very attractive due to several advantages such as first, high homogeneity of the film. Second, the force created by the vacuum pump helps the orientation of CNTs to be parallel to the surface, and therefore improves CNT to CNT electrical contacts. Last, the film thickness can be controlled by either varying the CNT solution concentration or varying the amount of solution that are filtered.

3.3.5. Ink-jet printing technique

CNT "inks" can be used to print thin CNT films on transparent substrate by using conventional bubble jet printers. One example of CNT ink is by mixing CNTs with water-soluble conducting polymer and poly(2-methoxyaniline-5-sulfonic acid) (PMAS) as shown by [43]. A

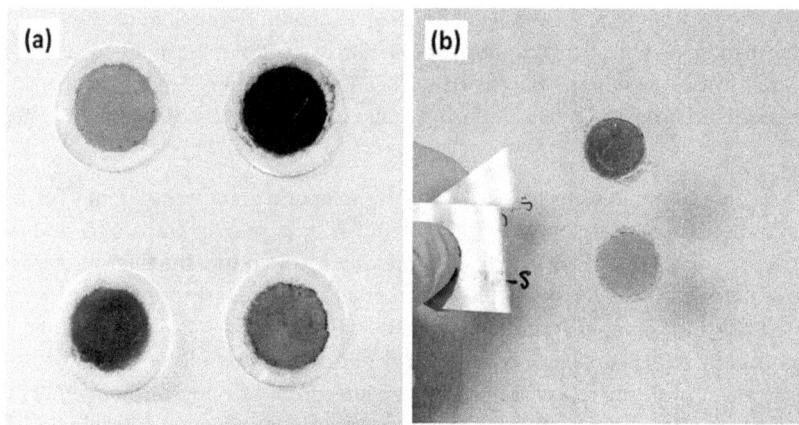

Figure 5. CNT films produced via vacuum filtration. (a) Membrane filters with CNT film of various thicknesses. (b) CNT transparent film after membrane removal and transfer on glass substrate.

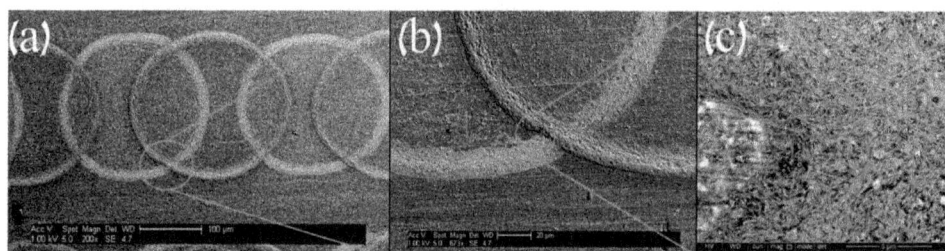

Figure 6. SEM images of connected ink-jet CNT droplet rings. Adapted from Ref. [44] with permission of The Royal Society of Chemistry.

single printed layer of CNT ink demonstrates a sheet resistance of ∼100 kΩ/□ (conductivity, σ, ∼0.93 S/cm) with transmittance of ∼85%. Interestingly, increasing the number of printed layers to four resulted in the reduction of sheet resistance by approximately a factor of two, but decreases the transmittance by no more than 10% [43].

It was also discovered that the ink-jet droplets can exhibit surface tension that can aggregate CNTs into bundles in the so-called coffee ring effect [44]. The surface tension of the ink-jet droplets created a uniform ring of CNT as thick as 300 nm that aggregate together as shown in **Figure 6**. By connecting each droplet ring to each other, a conducting network layer of CNT film has been produced by [44]. After temperature treatment, the resulting film exhibit sheet resistance as low as 1000 Ω/□, while the transmittance was ∼75%. The printed pattern of the network of CNT ring droplets can be adjusted to optimize the transparent conductor's performance.

3.3.6. Langmuir-Blodgett technique

Langmuir-Blodgett (LB) technique involves the self arrangement of organic material, including nanoparticles, forming a monolayer on the surface of a liquid in a trough. An LB film of

CNTs will be formed when there is an interaction at the water-air interface, whereby the naturally hydrophilic CNTs tend to aggregate at the surface. If the CNT concentration is less than the critical micellar concentration (CMC), the CNTs will arrange themselves as a mono-layer film parallel to the liquid surface. The density of the CNT LB film can be increased by pushing the liquid surface with a pedal-like instrument toward the center of the trough. Once a dense and aligned monolayer of CNT LB film is formed, a substrate can be immersed in the trough either parallel or perpendicular to the liquid surface where the film will then be absorbed onto the substrate's surface. The substrate immersion step is similar to the dip coating technique.

The CNT use in LB technique is usually dispersed and sonicated in organic compound with low boiling point and lower density than water such as chloroform. CNT dispersed in chloro-form is then carefully released on the surface of the liquid (typically deionized water) in the trough forming a monolayer of CNTs. Care must be taken to ensure that the density of CNT is less than the CMC. When the chloroform completely evaporates, the CNT LB film will remain suspended on the liquid surface ready to be compressed and transferred on a transparent substrate. To increase adhesion of the CNT LB film on the substrate, surface treatment with a primer or silane-based SAM.

A single layer of SWCNT LB film can have a thickness $\sim$3 nm, which is slightly higher than a typical SWCNT monolayer thickness due to the formation of bundles [45]. The LB process can be repeated to obtain thicker films. Since each LB process produces approximately monolayer CNT film, it possible to control the film thickness in the order of $\sim$3 nm. A SWCNT LB film of $\sim$300 nm thickness has been demonstrated by repeating the process 99 times [45]. In contrast, the dip coating technique does not offer tight control of CNT layer thickness after each subsequent dipping, and there is a limit to the maximum thickness that can be obtained depending on the adhesion or absorption sites that are available on the substrate's surface.

It is reported that a SWCNT LB film of $\sim$120 nm thick showed a transmittance of around 80–90% across the visible light spectrum [45]. This figure is comparable to the transmit-tance that of 20-nm thick film fabricated via the vacuum filtration method. **Figure 7** shows

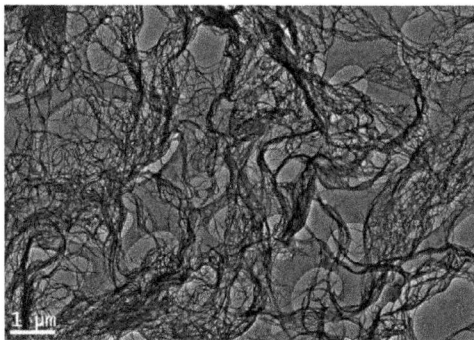

Figure 7. Transmission electron micrograph image of a 3-layer SWCNT LB film on an amorphous carbon grid. Adapted from Ref. [45] with permission from Elsevier.

a transmission electron microscope (TEM) image of a SWCNT LB film on carbon grid. The conductivity of the transparent film produced via the LB technique is dependent on the substrate orientation during the dipping step. Parallel direction of dipping showed the highest conductivity, followed by 45° angled and perpendicular directions, which showed the lowest value.

Although it is relatively easy to control the film thickness in the LB method, the electrical performance is generally inferior compared to other methods such as vacuum filtration and dip coating due to the poor intermolecular contact between adjacent CNTs in the film. This intermolecular contact can be improved by employing the hybrid film structure consisting of graphene oxide and SWCNTs via the LB method [46, 47]. This hybrid LB film exhibits the highest sheet resistance of 50,000 $\Omega/\square$ at a transmittance of ~97% and the lowest sheet resistance of 200 $\Omega/\square$ at 77% transmittance.

4. Comparison of transparent film performance

Generally, there exist a trade-off of CNT-based transparent conductor between the optical performance and the electrical performance. CNT transparent film with high transmittance exhibits reduced conductivity (or higher sheet resistance) and vice versa. However, there is another factor that affects the electrical performance of the film, which is the fabrication method or the film deposition method used. The inconsistency between the performances of the films produced by different methods may be due to the different intermolecular contact between adjacent CNTs in the film network. Discrepancy in the tube-tube contact resistance can result in significant collective conductivity of the whole film.

To study or quantify the tube-tube intermolecular contact resistances in CNT films from each fabrication method is difficult. One study has shown that in a controlled experiment, CNT transparent film fabrication via vacuum filtration method shows the best electrical conductivity, followed by dip coating and LB [48]. On the other hand, in terms of the transmittance, LB films show the highest transmittance, followed by dip coating and vacuum filtration. It is projected that a CNT transparent film fabricated via vacuum filtration with the same transmittance value with an LB film will show superior electrical conductivity due to the reduced intermolecular contact resistance between adjacent CNTs.

5. Conclusion

In this chapter, the basic theoretical background of carbon nanotube and its application as transparent film conductor have been discussed. The film fabrication methods available have also been discussed, along with the inherit advantages and disadvantages of each method. CNT-based transparent conductors have shown optical and electrical performance that rival that of ITO. Furthermore, CNT-based films can be applied on flexible substrates and do not suffer from temperature induced structural degradation as in other polymer- or organic-based films.

Acknowledgements

The authors would like to acknowledge the funding grants from The Ministry of Education (MoE), Malaysia: FRGS/1/2015/TK04/UKM/02/2; and Universiti Kebangsaan Malaysia (UKM): GP-K015333 and DIP-2016-021. The authors would also like to thank the Institute of Microengineering and Nanoelectronics (IMEN), Universiti Kebangsaan Malaysia (UKM), for supporting the work done in this chapter.

Conflict of interest

The author declares no conflict of interest.

Author details

Iskandar Yahya[1]*, Seri Mastura Mustaza[2] and Huda Abdullah[1]

*Address all correspondence to: iskandar.yahya@ukm.edu.my

1 Centre of Advanced Electronic and Communication Engineering (PAKET), Universiti Kebangsaan Malaysia (National University of Malaysia), Bangi, Selangor, Malaysia

2 Centre of (INTEGRA), Universiti Kebangsaan Malaysia (National University of Malaysia), Bangi, Selangor, Malaysia

References

[1] Farbod M, Zilaie A, Kazeminezhad I. Carbon nanotubes length optimization for preparation of improved transparent and conducting thin film substrates. Journal of Science-Advanced Materials and Devices. 2017;**2**(1):99-104

[2] He Y, Jin H, Qiu S, Li Q. A novel strategy for high-performance transparent conductive films based on double-walled carbon nanotubes. Chemical Communications. 2017;**53**(20): 2934-2937

[3] Tsapenko AP, Goldt AE, Shulga E, Popov ZI, Maslakov KI, Anisimov AS, et al. Highly conductive and transparent films of HAuCl4-doped single-walled carbon nanotubes for flexible applications. Carbon. 2018;**130**:448-457

[4] Yu Y, Luo Y, Guo A, Yan L, Wu Y, Jiang K, et al. Flexible and transparent strain sensors based on super-aligned carbon nanotube films. Nanoscale. 2017;**9**(20):6716-6723

[5] Jeon I, Yoon J, Ahn N, Atwa M, Delacou C, Anisimov A, et al. Carbon nanotubes versus graphene as flexible transparent electrodes in inverted perovskite solar cells. Journal of Physical Chemistry Letters. 2017;**8**(21):5395-5401

[6] Pyo S, Kim W, Jung H, Choi J, Kim J. Heterogeneous Integration of Carbon-Nanotube-Graphene for High-Performance, Flexible, and Transparent Photodetectors. Small. 2017;**13**(27)

[7] Wildoer JWG, Venema LC, Rinzler AG, Smalley RE, Dekker C. Electronic structure of atomically resolved carbon nanotubes. Nature. 1998;**391**(6662):59-62

[8] Thess A, Lee R, Nikolaev P, Dai H, Petit P, Robert J, et al. Crystalline ropes of metallic carbon nanotubes. Science. 1996;**273**(5274):483-487

[9] Avouris P. Carbon nanotube electronics. Chemical Physics. 2002;**281**(2-3):429-445

[10] White C, Todorov T. Carbon nanotubes as long ballistic conductors. Nature. 1998;**393**(6682): 240-242

[11] Hone J, Whitney M, Zettl A. Thermal conductivity of single-walled carbon nanotubes. Synthetic Metals. 1999;**103**(1-3):2498-2499

[12] Wong EW, Sheehan PE, Lieber CM. Nanobeam mechanics: Elasticity, strength, and toughness of nanorods and nanotubes. Science. 1997;**277**(5334):1971-1975

[13] Yu MF, Lourie O, Dyer MJ, Moloni K, Kelly TF, Ruoff RS. Strength and breaking mechanism of multi-walled carbon nanotubes under tensile load. Science. 2000;**287**(5453):637-640

[14] Schonenberger C, Bachtold A, Strunk C, Salvetat JP, Forro L. Interference and Interaction in multi-wall carbon nanotubes. Applied Physics A: Materials Science and Processing. 1999;**69**(3):283-295

[15] de Heer WA, Martel R. Industry sizes up nanotubes. Physics World. 2000;**13**(6):49-53

[16] Ebbesen TW, Ajayan PM. Large-scale synthesis of carbon nanotubes. Nature. 1992;**358**(6383): 220-222

[17] Dai HJ, Kong J, Zhou CW, Franklin N, Tombler T, Cassell A, et al. Controlled chemical routes to nanotube architectures, physics, and devices. The Journal of Physical Chemistry. B. 1999; **103**(51):11246-11255

[18] Baker RTK. CATALYTIC GROWTH OF CARBON FILAMENTS. Carbon. 1989;**27**(3):315-323

[19] Li YM, Mann D, Rolandi M, Kim W, Ural A, Hung S, et al. Preferential growth of semiconducting single-walled carbon nanotubes by a plasma enhanced CVD method. Nano Letters. 2004;**4**(2):317-321

[20] Nikolaev P, Bronikowski MJ, Bradley RK, Rohmund F, Colbert DT, Smith KA, et al. Gas-phase catalytic growth of single-walled carbon nanotubes from carbon monoxide. Chemical Physics Letters. 1999;**313**(1-2):91-97

[21] Kitiyanan B, Alvarez WE, Harwell JH, Resasco DE. Controlled production of single-wall carbon nanotubes by catalytic decomposition of CO on bimetallic Co-Mo catalysts. Chemical Physics Letters. 2000;**317**(3-5):497-503

[22] Maruyama S, Kojima R, Miyauchi Y, Chiashi S, Kohno M. Low-temperature synthesis of high-purity single-walled carbon nanotubes from alcohol. Chemical Physics Letters. 2002; **360**(3-4):229-234

[23] Hecht D, Hu L, Irvin G. Emerging transparent electrodes based on thin films of carbon nanotubes, graphene, and metallic nanostructures. Advanced Materials. 2011;**23**(13): 1482-1513

[24] Zhao M-Q, Tian G-L, Zhang Q, Huang J-Q, Nie J-Q, Wei F. Preferential growth of short aligned, metallic-rich single-walled carbon nanotubes from perpendicular layered double hydroxide film. Nanoscale. 2012;**4**(7):2470-2477

[25] Li J, Liu K, Liang S, Zhou W, Pierce M, Wang F, et al. Growth of high-density-aligned and semiconducting-enriched single-walled carbon nanotubes: Decoupling the conflict between density and selectivity. ACS Nano. 2014;**8**(1):554-562

[26] Krupke R, Hennrich F, von Lohneysen H, Kappes MM. Separation of metallic from semiconducting single-walled carbon nanotubes. Science. 2003;**301**(5631):344-347

[27] Mattsson M, Gromov A, Dittmer S, Eriksson E, Nerushev OA, Campbell EEB. Dielectro-phoresis-induced separation of metallic and semiconducting single-wall carbon nanotubes in a continuous flow microfluidic system. Journal of Nanoscience and Nanotechnology. 2007;**7**(10):3431-3435

[28] Tanaka T, Jin HH, Miyata Y, Kataura H. High-yield separation of metallic and semicon-ducting single-wall carbon nanotubes by agarose gel electrophoresis. Applied Physics Express. 2008;**1**(11)

[29] Liu H, Nishide D, Tanaka T, Kataura H. Large-scale single-chirality separation of single-wall carbon nanotubes by simple gel chromatography. Nature Communications. 2011;**2**:309

[30] Yahya I, Bonaccorso F, Clowes SK, Ferrari AC, Silva SRP. Temperature dependent separa-tion of metallic and semiconducting carbon nanotubes using gel agarose chromatography. Carbon. 2015;**93**:574-594

[31] Arnold MS, Green AA, Hulvat JF, Stupp SI, Hersam MC. Sorting carbon nanotubes by electronic structure using density differentiation. Nature Nanotechnology. 2006;**1**(1): 60-65

[32] Collins PG, Hersam M, Arnold M, Martel R, Avouris P. Current saturation and electrical breakdown in multi-walled carbon nanotubes. Physical Review Letters. 2001;**86**(14): 3128-3131

[33] Wu ZC, Chen ZH, Du X, Logan JM, Sippel J, Nikolou M, et al. Transparent, conductive carbon nanotube films. Science. 2004;**305**(5688):1273-1276

[34] Zhang M, Fang SL, Zakhidov AA, Lee SB, Aliev AE, Williams CD, et al. Strong, transparent, multifunctional, carbon nanotube sheets. Science. 2005;**309**(5738):1215-1219

[35] Ding L, Yuan DN, Liu J. Growth of high-density parallel arrays of long single-walled carbon nanotubes on quartz substrates. Journal of the American Chemical Society. 2008;**130**(16):5428

[36] Shi D, Resasco DE. Study of the growth of conductive single-wall carbon nanotube films with ultra-high transparency. Chemical Physics Letters. 2011;**511**(4–6):356-362

[37] Duc Dung N, Tai N-H, Chen S-Y, Chueh Y-L. Controlled growth of carbon nanotube-graphene hybrid materials for flexible and transparent conductors and electron field emitters. Nanoscale. 2012;**4**(2):632-638

[38] Hecht DS, Heintz AM, Lee R, Hu L, Moore B, Cucksey C, et al. High conductivity transparent carbon nanotube films deposited from superacid. Nanotechnology. 2011;**22**(7)

[39] Rastogi R, Kaushal R, Tripathi SK, Sharma AL, Kaur I, Bharadwaj LM. Comparative study of carbon nanotube dispersion using surfactants. Journal of Colloid and Interface Science. 2008;**328**(2):421-428

[40] Huang WJ, Lin Y, Taylor S, Gaillard J, Rao AM, Sun YP. Sonication-assisted functionalization and solubilization of carbon nanotubes. Nano Letters. 2002;**2**(3):231-234

[41] Song YI, Yang C-M, Kim DY, Kanoh H, Kaneko K. Flexible transparent conducting single-wall carbon nanotube film with network bridging method. Journal of Colloid and Interface Science. 2008;**318**(2):365-371

[42] Kim S, Yim J, Wang X, Bradley DDC, Lee S, de Mello JC. Spin- and spray-deposited single-walled carbon-nanotube electrodes for organic solar cells. Advanced Functional Materials. 2010;**20**(14):2310-2316

[43] Small WR, Panhuis MIH. Inkjet printing of transparent, electrically conducting single-waited carbon-nanotube composites. Small. 2007;**3**(9):1500-1503

[44] Shimoni A, Azoubel S, Magdassi S. Inkjet printing of flexible high-performance carbon nanotube transparent conductive films by "coffee ring effect". Nanoscale. 2014;**6**(19):11084-11089

[45] Massey MK, Pearson C, Zeze DA, Mendis BG, Petty MC. The electrical and optical properties of oriented Langmuir-Blodgett films of single-walled carbon nanotubes. Carbon. 2011;**49**(7):2424-2430

[46] Zheng Q, Zhang B, Lin X, Shen X, Yousefi N, Huang Z-D, et al. Highly transparent and conducting ultralarge graphene oxide/single-walled carbon nanotube hybrid films produced by Langmuir-Blodgett assembly. Journal of Materials Chemistry. 2012;**22**(48):25072-25082

[47] Yang T, Yang J, Shi L, Maeder E, Zheng Q. Highly flexible transparent conductive graphene/single-walled carbon nanotube nanocomposite films produced by Langmuir-Blodgett assembly. RSC Advances. 2015;**5**(30):23650-23657

[48] Yahya I, Theng L, Mustaza S, Abdullah H, Amin N. Characterization of transparent conducting carbon nanotube thin films prepared via different methods. Sains Malaysiana. 2017;**46**(7):1103-1109

Application of Chemically Synthesized Conductive Polymers for Biosensing

Cyclic Voltammetry and Electrical Impedance Spectroscopy of Electrodes Modified with PEDOT:PSS-Reduced Graphene Oxide Composite

Nurul Izzati Ramli, Nur Alya Batrisya Ismail,
Firdaus Abd-Wahab and
Wan Wardatul Amani Wan Salim

Additional information is available at the end of the chapter

http://dx.doi.org/10.5772/intechopen.80715

Abstract

Cyclic voltammetry (CV) and electrical impedance spectroscopy (EIS) are electrochemical techniques to characterize reversibility of electron transfer and impedance at the electrode-solution interface, respectively. Reduced graphene oxide (rGO) and conductive polymer PEDOT:PSS are often used to enhance electron transfer at electrode surface. This chapter provides a step-by-step methodology of CV and EIS conducted on screen-printed carbon electrode (SPCE) modified with rGO-PEDOT:PSS and brief discussion on the CV and EIS results. The CV of rGO-PEDOT:PSS shows a reversible electron transfer in comparison to SPCE modified with PEDOT:PSS. For EIS, rGO-PEDOT:PSS was found to reduce the Warburg impedance, suggesting enhanced electrode conductivity. These results suggest that rGO-PEDOT:PSS is a suitable material for biosensing purpose.

Keywords: PEDOT:PSS, graphene, cyclic voltammetry, electrical impedance spectroscopy, biosensor

1. Introduction

A polymer is made up of repeating subunits of monomers and widely recognized as having good insulating properties. In 1977, three scientists serendipitously discovered that polyacetylene (PA) could become conductive through iodine doping, which consequently allows electrons at the π-bonds to move along the polymeric chains. This process enables the development of conductive polymers and their exploitation in many applications such as supercapacitors, light-emitting diodes (LEDs), solar cells, field-effect transistors (FETs), and biosensors [1, 2].

One such conductive polymer that is gaining interest is the poly(3,4-ethylenedioxythio phene):poly(styrene sulfonate) or PEDOT:PSS. PEDOT:PSS exhibits high conductivity [3], high stability in liquid media [4], and good mechanical flexibility [5]. These properties make PEDOT:PSS a suitable transducer material in electrochemical biosensors. Furthermore, its ability to intercalate between the layers of reduced graphene oxide (rGO) opens the possibility of combining these two materials as a composite to enhance biosensor performance.

This chapter describes the experimental procedures for the use of PEDOT:PSS and rGO as the transducer composite on a screen-printed carbon electrode (SPCE). Electrochemical analyses are done via cyclic voltammetry (CV) and electrical impedance spectroscopy (EIS) to characterize the electrochemical reversibility of the material, as well as its impedance at the electrode-solution interface.

1.1. Cyclic voltammetry (CV)

Cyclic voltammetry (CV) is used to understand and characterize the redox characteristics, stability, and effective surface area of an electrode for biosensing; such electrodes, modified with materials such as PEDOT:PSS and graphene, function as a transducer to convert ions into measurable electrons. CV typically consists of a three-electrode setup of a working electrode (WE), a reference electrode (RE), and a counter electrode (CE). In CV, a potential applied to the WE is swept back and forth for a defined number of cycles over a given range of voltage and speed of voltage sweep [6]. As the potential is scanned across a specified potential range, the resulting current at the WE is measured. The current generated is then plotted against potential to produce a CV graph that provides insights on the transducer material based on the anodic peak current (I_{pa}) from oxidation process and cathodic peak current (I_{pc}) from reduction process which occur on the WE. The potentials at which the peak currents occur are known as peak potentials (E_p). These peak potentials enable us to analyze the electrochemical reversibility of the reaction at the electrode surface by increasing the scan rates during experiments. Electrochemically reversible reactions often establish fast electron transfer between species and electrode, whereas the electron transfer in irreversible reaction is slower [6, 7].

1.1.1. Experiment

Figure 1 shows the experimental setup for CV using a SPCE connected to a potentiometer and a laptop.

CV experiments were conducted on SPCEs modified with PEDOT:PSS and rGO-PEDOT:PSS composite.

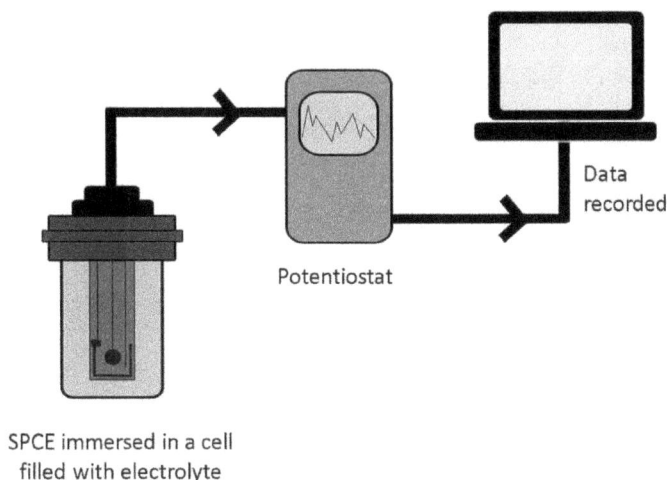

Figure 1. Experimental setup for CV. The SPCE is immersed in a cell filled with electrolyte. The cell is connected to a potentiometer and a laptop.

1.1.2. Materials and equipment

1.1.2.1. Consumable items and equipment

- Magnetic stirrer

- Sonicator

- Pipette and pipette tips

- Potentiometer (IVIUM Technologies, Eindhoven, Netherlands)

- Screen-printed carbon electrode (SPCE with 2 mm diameter, Pine Instruments, Grove City, Pennsylvania, USA)

1.1.2.2. Chemicals and reagents

- PEDOT:PSS solution (1.3 wt% dispersion in H_2O, high-conductivity grade, Sigma-Aldrich, St. Louis, MO, USA)

- Ultrahighly concentrated GO (UHC GO) solution (6.2 mg/ml, Graphene Supermarket, USA)

- M potassium ferricyanide $K_3Fe(CN)_6$ (R&M Chemicals, Selangor, Malaysia)

- M PBS, pH 5 (Sigma-Aldrich, St. Louis, MO, USA)

- Deionized (DI) water

1.1.3. Method

1.1.3.1. Fabrication of PEDOT:PSS/SPCE

1. Drop cast 3 μl PEDOT:PSS onto the WE of a SPCE.

2. Dry for 24 h in ambient conditions.

1.1.3.2. Fabrication of rGO-PEDOT:PSS/SPCE

1. Mix 500 μl PEDOT:PSS with 500 μl ultrahighly concentrated GO solution.

2. Stir the solution using a magnetic stirrer, and sonicate at 30°C for 10 min in order to form a well-mixed GO and PEDOT:PSS solution.

3. Drop cast 3 μl solution onto the WE of a SPCE, and dry in ambient conditions.

4. Reduce the electrode using a portable potentiostat via repetitive cyclic voltammetry (CV) with potential range from 0 to −1.5 V at 0.1 V/s in 0.01 M PBS, pH 5, for 15 cycles. Make sure that all electrodes (WE, RE, and CE) are fully dipped in the PBS during the reduction process. The reduced electrode is denoted as rGO-PEDOT:PSS/SPCE.

5. Rinse the rGO-PEDOT:PSS/SPCE with DI water, and dry at room temperature for 24 h.

1.1.3.3. CV measurement

1. To perform CV on the rGO-PEDOT:PSS/SPCE, insert the modified SPCE into a cell filled with 0.1 M $K_3Fe(CN)_6$ solution. Make sure all electrodes (WE, RE, and CE) are immersed in the solution.

2. Run a CV scan from initial potential of −0.5 to 1 V and back to −0.5 V at a scan rate of 25 mV/s. Repeat the scans at 50, 100, 150, and 200 mV/s.

1.1.4. Results

1.1.4.1. Discussion of results

Figures 2 and **3** show CV plots for PEDOT:PSS and rGO-PEDOT:PSS in 0.1 M $K_3Fe(CN)_6$ with increasing scan rates of 25, 50, 100, 150, and 200 mV/s. The peak current (I_p) produced at 200 mV/s is higher than the I_p at 25 mV/s for both PEDOT:PSS/SPCE and rGO-PEDOT:PSS/SPCE. However, the peak potentials (E_p) of the rGO-PEDOT:PSS transducer are shifted to more extreme values with increased scan rate than PEDOT:PSS transducer. The shift in the E_p as observed in the PEDOT:PSS/SPCE suggests that the electron transfer takes place heterogeneously—the redox species adsorbs onto the electrode surface prior to electron transfer. The addition of rGO to the composite could help reduce the adsorption of the redox species and enable the electron to be electrochemically reversible [6, 7]. This also suggests that

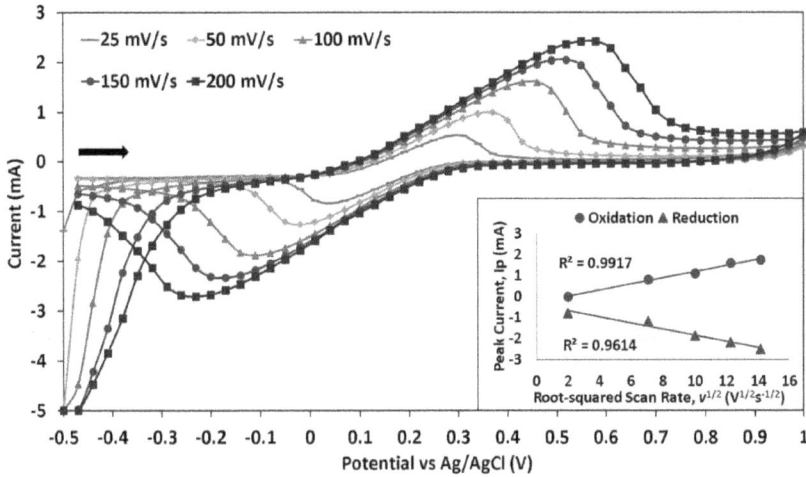

Figure 2. Cyclic voltammetry at different scan rates for PEDOT:PSS transducer. The electrolyte used in this experiment is 0.1 M $K_3Fe(CN)_6$ solution. Voltage was swept at potentials from −0.5 to 1 V and back. The arrow shows the direction of the initial voltage sweep. Inset shows the Cottrell plot for the CV.

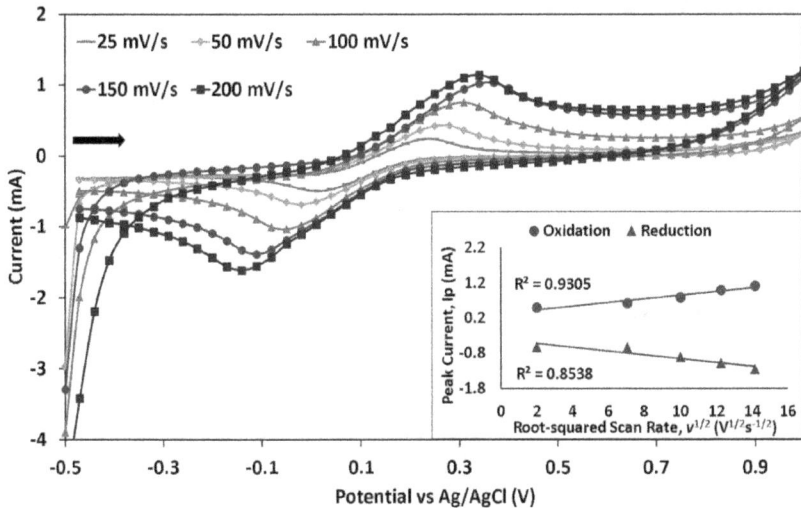

Figure 3. Cyclic voltammetry at different scan rates for an electrode modified with reduced graphene oxide and PEDOT:PSS as the transducer. The electrolyte used in this experiment is 0.1 M $K_3Fe(CN)_6$. Voltage was swept at potentials from −0.5 to 1 V and back. The arrow shows the direction of the initial voltage sweep. Inset shows the Cottrell plot for the CV.

PEDOT:PSS with rGO can potentially increase the electron transfer rate at the electrodes, further improving the performance of the biosensor in terms of sensitivity to changes in analyte concentration.

1.2. Electrical impedance spectroscopy (EIS)

EIS is an analytical tool to study the interfacial behavior occurring on the surface of an electrode. The impedance of an electrode at the electrode-solution interface is determined by applying a small alternating (AC) sinusoidal voltage (~10 mV peak-to-peak) perturbation and tracking the current output. Since an AC is applied across the surface of the electrode, the voltage–current output will also be observed in a range of frequencies [8]. The experimental data can be represented in two ways, i.e., the Nyquist plot (where the real vs. imaginary impedance components are plotted) and the Bode plot (where the impedance and phase angle were plotted against frequency). The data obtained from experimental studies can be analyzed and evaluated using an equivalent circuit comprising a series of resistances and capacitances in parallel (Randles equivalent circuit) [9]. Note that the Nyquist plot will be preferred in this study because of the extensive information that can be obtained to understand the impedance at the electrode-solution interface, charge transfer resistance, and Warburg impedance.

1.2.1. Experiments

A similar experimental setup as **Figure 1** was used for EIS measurements. The screen-printed carbon electrode (SPCE) was connected to a potentiometer, and the signal obtained was processed for further analysis.

1.2.2. Materials

1.2.2.1. Consumable items and equipment

- Magnetic stirrer

- Sonicator

- Pipette and pipette tips

- Potentiometer (IVIUM Technologies, Eindhoven, Netherlands)

- Screen-printed glassy electrode (SPCE with 2 mm diameter, Pine Instruments, Grove City, Pennsylvania, USA)

1.2.2.2. Chemicals and reagents

- PEDOT:PSS solution (1.3 wt% dispersion in H_2O, high-conductivity grade, Sigma-Aldrich, St. Louis, MO, USA)

- Ultrahighly concentrated GO (UHC GO) solution (6.2 mg/ml, Graphene Supermarket, USA)

- 5 mM potassium ferricyanide/potassium ferrocyanide $K_3Fe(CN)_6/K_4Fe(CN)_6$ in 0.01 M PBS, pH 7.4 (R&M Chemicals, Selangor, Malaysia)

- M PBS, pH 5 (Sigma-Aldrich, St. Louis, MO, USA)

- Deionized (DI) water

1.2.3. Methods

1.2.3.1. Fabrication of rGO-PEDOT:PSS electrode

1. Mix 500 μl PEDOT:PSS with 500 μl UHC GO solution (1,1) in a small vial.

2. Stir the solution using a magnetic stirrer, and sonicate at 30°C for 10 min to ensure homogenous mixture of GO-PEDOT:PSS nanocomposite.

3. Drop cast 3 μl GO-PEDOT:PSS mixture onto the WE of a SPCE, and leave to dry at ambient temperature.

4. Reduce the electrode using a potentiostat via repetitive cyclic voltammetry (CV) with potential range from 0 to −1.5 V at 0.1 V/s in 0.01 M PBS, pH 5, for 15 cycles. Ensure that all electrodes (WE, RE, and CE) are fully immersed in the PBS during the reduction process. The reduced electrode is denoted as rGO-PEDOT:PSS/SPCE.

5. Rinse the rGO-PEDOT:PSS/SPCE with DI water, and dry at ambient temperature for 2 h.

1.2.3.2. EIS measurement

1. To perform EIS measurement, dip the rGO-PEDOT:PSS/SPCE into a cell filled with 5 mM $K_3Fe(CN)_6/K_4Fe(CN)_6$ (1:1) in 0.01 M PBS, pH 7.4. Make sure that all electrodes (WE, RE, and CE) are immersed in the solution.

2. Run EIS over a frequency range of 100 kHz–0.1 Hz, amplitude of 20 mV, and applied potential of 0.25 V.

3. Observe and record the real impedance (Z′) vs. imaginary impedance (−Z″) obtained from the EIS measurement. Repeat steps 1–4 with a bare SPCE electrode to show the comparison between SPCE and rGO-PEDOT:PSS/SPCE.

1.2.4. Results

Figure 4 shows a Nyquist plot of real impedance (Z′) vs. imaginary impedance (−Z″) for a SPCE and an rGO-PEDOT:PSS/SPCE. The observed plots are fitted to a Randles equivalent circuit for analysis. A typical Randles circuit includes the solution resistance (R_s), charge transfer resistance (R_{ct}), Warburg impedance (Z_w), and double-layer capacitance (C_{dl}). R_s refers to the resistance between a solution containing ions at a certain concentration and an electrochemical cell, R_{ct} refers to the electron transfer produced by a redox reaction at the electrode interface, Z_w indicates the impedance of electrons due to the diffusion interface between bulk

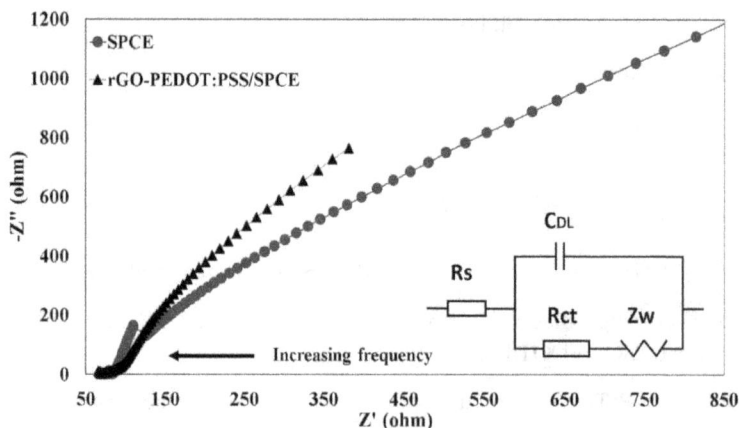

Figure 4. EIS of SPCE and rGO-PEDOT:PSS/SPCE in PBS, pH 7.4, containing 5 mM $K_3[Fe(CN)_6]/K_4[Fe(CN)_6]$ (1:1) over a frequency range of 100 kHz–0.1 Hz, amplitude of 20 mV, and applied potential of 0.25 V. Inset shows Randles equivalent circuit of a typical biological sensor.

solution and the electrode interface, and C_{dl} is the capacitance obtained from stored charge in the double-layer region at the electrode interface [10, 11].

1.2.4.1. Discussion of results

The Nyquist graph in **Figure 4** shows impedance measurements of SPCE and rGO-PEDOT:PSS/ SPCE. Since impedance is a measure of resistance occurring on the electrode interface, the Nyquist plots show that the impedance for rGO-PEDOT:PSS/SPCE increases in comparison to a bare SPCE. In a typical Nyquist plot, a semicircular region is observed in the lower-frequency region, whereas a linear plot is observed in the high-frequency region. For a bare SPCE, the plots are dominated by the linear region, which indicates that the electron transfer occurring on the electrode surface is heavily influenced by a diffusion-limited process [12]. The diffusion-limited process is typically represented by the Warburg impedance (Z_w), which means that the impedance is a result of the diffusion layer between the bulk solution and the electrode surface. For the rGO-PEDOT:PSS/SPCE, the linear region is slightly shorter than that of the bare SPCE, and there is a small half semicircle observed at the low-frequency region, which suggests that the influence of the Warburg impedance is low for the rGO-PEDOT:PSS/ SPCE [8]. The results also suggest that the electrode shows a slightly capacitive behavior, which could be owing to the presence of PEDOT:PSS in the electrode interface [4, 13].

2. Conclusion

Voltammetry characterization of CV and EIS for a conducting rGO-PEDOT:PSS composite is presented. Both CV and EIS are relevant techniques for understanding electron transfer at the electrode-solution interface. The results from CV indicate that the presence of rGO in rGO-PEDOT:PSS/SPCE helps to enable electrochemically reversible electron transfer in comparison to PEDOT:PSS/SPCE by reducing the shift in the peak potential (E_p). For EIS, the rGO-PEDOT:PSS/SPCE reduces the influence of Warburg impedance in comparison to a bare SPCE. These findings suggest that the rGO-PEDOT:PSS composites in electrochemical devices are very suitable for application in biological sensing.

Acknowledgements

The author would like to acknowledge the Malaysian Ministry of Education for awarding the financial support via the Research Initiative Grant Scheme RIGS16-355-0519 and RIGS Postdoctoral RPDF18-002-0002.

Conflict of interest

The authors declare no conflict of interest.

Abbreviations

CE	counter electrode
CV	cyclic voltammetry
EIS	electrical impedance spectroscopy
PBS	phosphate-buffered saline
PEDOT:PSS	poly(3,4-ethylenedioxythiophene):poly(styrene sulfonate)
rGO	reduced graphene oxide
RE	reference electrode
SPCE	screen-printed carbon electrode
WE	working electrode

Author details

Nurul Izzati Ramli, Nur Alya Batrisya Ismail, Firdaus Abd-Wahab
and Wan Wardatul Amani Wan Salim*

*Address all correspondence to: asalim@iium.edu.my

Department of Biotechnology Engineering, Faculty of Engineering, International Islamic University Malaysia (IIUM), Kuala Lumpur, Malaysia

References

[1] Atesa M, Karazehir T, Sarac AS. Conducting polymers and their applications. Current Physical Chemistry. 2012;**2**(3):224-240

[2] Park CS, Lee C, Kwon OS. Conducting polymer based nanobiosensors. Polymers. 2016;**8**(7):1-18

[3] Ouyang J, Chu CW, Chen FC, Xu Q, Yang Y. High-conductivity poly(3,4-ethylenediox ythiophene):poly(styrene sulfonate) film and its application in polymer optoelectronic devices. Advanced Functional Materials. 2005;**15**(2):203-208

[4] Benoudjit A, Bader MM, Wan Salim WWA, Salim WWAW. Study of electropolymerized PEDOT:PSS transducers for application as electrochemical sensors in aqueous media. Sensing and Bio-Sensing Research. 2018;**17**(1):18-24. DOI: 10.1016/j.sbsr.2018.01.001

[5] Wang Z, Xu J, Yao Y, Zhang L, Wen Y, Song H, et al. Facile preparation of highly water-stable and flexible PEDOT:PSS organic/inorganic composite materials and their application in electrochemical sensors. Sensors and Actuators B: Chemical. 2014;**196**:357-369

[6] Elgrishi N, Rountree KJ, McCarthy BD, Rountree ES, Eisenhart TT, Dempsey JL. A practical beginner's guide to cyclic voltammetry. Journal of Chemical Education. 2018;**95**(2):197-206

[7] Graham D. Testing for Adsorption [Internet]. Available from: https://sop4cv.com/chapters/TestingForAdsorption.html [Cited: July 20, 2018]

[8] Santos A, Davis JJ, Bueno PR. Fundamentals and applications of impedimetric and redox capacitive biosensors. Journal of Analytical and Bioanalytical Techniques. 2014;**S7**(016):1-15. Available from: https://www.omicsonline.org/open-access/fundamentals-and-applications-of-impedimetric-and-redox-capacitive-biosensors-2155-9872.S7-016.php?aid=28343

[9] Randviir EP, Banks CE. Electrochemical impedance spectroscopy: An overview of bio-analytical applications. Analytical Methods. 2013;**5**(5):1098. Available from: http://xlink.rsc.org/?DOI=c3ay26476a

[10] Lisdat F, Schäfer D. The use of electrochemical impedance spectroscopy for biosensing. Analytical and Bioanalytical Chemistry. 2008;**391**(5):1555-1567

[11] Orazem ME, Tribollet B. Electrochemical Impedance Spectroscopy. Hoboken New Jersey, USA: John Wiley & Sons Inc.; 2008. 518 p

[12] Fernández-Sánchez C, McNeil CJ, Rawson K. Electrochemical impedance spectroscopy studies of polymer degradation: Application to biosensor development. TrAC Trends in Analytical Chemistry. 2005;**24**(1):37-48

[13] Fernandes FCB, Góes MS, Davis JJ, Bueno PR. Label free redox capacitive biosensing. Biosens Bioelectron. 2013;**50**:437-440. DOI: 10.1016/j.bios.2013.06.043

www.ingramcontent.com/pod-product-compliance
Lightning Source LLC
Chambersburg PA
CBHW070241230326
41458CB00100B/5812